THE ENDS OF RACE

The Ends of Race

THE RISE AND FALL OF A SCIENTIFIC MYTH

Geoffrey Galt Harpham

PRINCETON UNIVERSITY PRESS
PRINCETON & OXFORD

Published by Princeton University Press
41 William Street, Princeton, New Jersey 08540
99 Banbury Road, Oxford OX2 6JX

press.princeton.edu

GPSR Authorized Representative: Easy Access System Europe - Mustamäe tee 50, 10621 Tallinn, Estonia, gpsr.requests@easproject.com

ISBN 978-0-691-29162-8
ISBN (epub) 978-0-691-29559-6
ISBN (PDF) 978-0-691-29163-5

British Library Cataloging-in-Publication Data is available

Editorial: Ben Tate and Josh Drake
Production Editorial: Jill Harris
Jacket Design: Haley Jin Mee Chung
Production: Danielle Capurso
Publicity: Alyssa Sanford and Charlotte Coyne
Copyeditor: Cindy Milstein

Jacket image: Paper texture by Shutterstock

This book has been composed in Miller

Printed in the United States of America

10 9 8 7 6 5 4 3 2 1

MIX
Paper
FSC FSC® C183721

For Bernard Lategan, theologian and founding director, Stellenbosch Institute for Advanced Study

CONTENTS

PREFACE

SO THOROUGHLY EMBEDDED in Western ways of thinking is race that many people find it difficult to understand that the concept itself has a history. But it does. It was developed by people who believed that an empirical inquiry into the origins, depth, and character of certain divisions within the category of the human might result in a major scientific triumph and, just as important, might have certain desirable worldly consequences. Even more difficult to fully appreciate is the fact that the scientific effort to identify a system of biologically determined categories collapsed over a century ago, so that we are now living through the ghostly afterlife of a discredited project, with the concept sustained now as then not by the evidence that supports it—there is none—but by the ends it supports. But this too is true.

This book provides a skeletal version of the history of the concept of race from its uncertain beginnings in the late eighteenth century to the end of the nineteenth. It is skeletal in two primary senses. First, it is a reduction that leaves out a great deal in order to reveal the essential form of the whole. Race was so thoroughly entangled with other concepts and worldly forces during this time that a complete account would involve a multinational study of philosophy, history, literature, scientific methodology, theology, geography, anthropology, economics—really, it would involve everything. Clarity and coherence require selection, so I have concentrated on a few critical moments or controversies in which the most representative and consequential issues associated with the concept of race were thrashed out.

Within a teeming field of possibility, I trace a linear narrative that begins in Germany, moves to the United States, and ends in France, with one overarching question addressed in each setting. The result, while partial in that significant developments in nineteenth-century Germany and Britain are given very short shrift, is, I think, revealing in ways that a more comprehensive or fleshed-out account could not be. Race itself is often described as being both real and "constructed," and that is how the story told in this book should be considered.

This account is also skeletal in that the individual chapters are connected by bones. Throughout the period canvased here, scientists and intellectuals trying to discover the elusive secret of human variation kept returning to the possibility that hard evidence was to be found not in skin color, hair texture, or blood, but in bony tissue, especially the skull—"the hardest possible receptacle," as Johann Friedrich Blumenbach put it, of "the softest and noblest entrail." Throughout the nineteenth century, many leading thinkers came to suspect or even to believe that patterns in skull shapes and sizes could provide clues about the categorical organization of humanity, with boundary lines of some kind but natural in origin separating the groups. Craniology, which now seems an aberration or even a perversion of science, commanded more respect in the intellectual communities of Western nations than any other approach to the problem of human variation in large part because it was so manifestly "scientific," that is, empirical, positivistic, and statistical.

There is also a third sense in which the tale told here can be considered skeletal. The attempts to theorize race during this period have acquired the character of skeletons in the closet of Western intellectual and cultural history. The title of one of the most informative treatments of the subject, C. Loring Brace's *"Race" Is a Four-Letter Word*, captures the sense that the entire project of developing a concept of race was morally as well as intellectually disgraceful, a prolonged and motivated blunder, or even an unholy compact in which scholarship and science served as high-prestige but bad-faith enablers of power. For many recent thinkers, the theorizing of race during this time represents an episode in the history of Western arrogance toward non-Westerners, a history that extends as far back as ancient Egypt and continues to the present day, often operating in disguised or denied forms.

The sinister implications and entailments of the concept of race that now seem inescapable were, however, far from the minds of most of those who first explored and promoted the concept in the late eighteenth century. For them, a novel and somewhat blurred term applied to royal families, professions, and horses offered itself as a way of raising questions to which neither the Bible nor philosophical reflection had been able to provide explanations satisfying to a modern intellectual culture. An indefinite but suggestive term, race hovered between species, a category within the natural order that scientists were beginning to map, and other terms such as variety, nation, people, or stock that referred to populations shaped by external factors such as climate, geography, and culture.

The conspicuous advantage of race as a heading for inquiry was that the term was not decisively natural or cultural but was ambiguously both, and so could be used by thinkers of any persuasion who wanted to inquire into a subject widely recognized as having potentially great importance for an understanding of the human condition: the origin, character, purpose, depth, and consequence of difference.

The uncertain character of the term permitted a good deal of latitude on the part of those who used it. For some including Blumenbach, the term was of doubtful value in a scientific context; for others such as James Cowles Prichard and Paul Broca, ambiguity was a positive feature, allowing research and inquiry to proceed even before the larger question of whether anything in nature corresponded to the concept of race had been settled. For many others including Christoph Meiners and Josiah C. Nott, there was no ambiguity, and the meaning of race was clear: certain groups, which they called races, were so different from each other as to be virtually separate species, and "the human race" was a misleading name for a nonexistent totality. And for Frederick Douglass and others especially at the end of the nineteenth century, the term itself was dangerous because it suggested a false principle of division and should not be applied to anything other than to humanity as a whole. A promiscuous concept, race could be deployed in the service of many different causes—the "ends of race" of my title.

As the nineteenth century progressed and race came commandingly to the fore as a subject for scientific research, the term acquired associations with, and implications for, a host of historical, political, and moral issues, many of them centering on practices of colonization, exploitation, or enslavement. These associations took on the aspect of a penumbra, giving a polemical and proxy character to even the most limited inquiries or research projects. To think of race was to think either implicitly or explicitly, consciously or unconsciously, of things other than race. The real stakes of studies of hybridity, taxonomic charts, racial maps, statistics on the internal capacity of skulls, or mortality figures for different populations vastly exceeded the subjects ostensibly under review, giving the discourse on race during this period a distinctive sense of dimension, energy, and importance.

Lacking a precise definition of the object and infused with extrascientific concerns, the study of race did not advance in the manner of what might be called normal science, by systematic inquiry undertaken by disinterested scholars or researchers who agreed on the object of study, the appropriate methodology, and the rules of evidence. Nor did it take the form of isolated statements made by individuals on clearly defined subjects. The concept was hammered out in agonistic fashion through a series of sprawling and often eruptive confrontations between sharply etched combatants flinging out pronouncements and declarations into a high wind of opposition on a range of subjects, of which race was one and often not the primary one.

In retrospect, the most compelling feature of these confrontations was not the problems solved or the debates won or lost. No problems were actually solved, and in that respect everyone lost. The real interest for us lies in the way in which the confrontations themselves revealed the mind of the disputants and the state of the wider culture. Because the evidence for race was so speculative and contested, and because the potential implications of the inquiry were so profound, the discourse on the subject often had an unmasking effect on the premises one began with, the arguments one found compelling, the evidence one was persuaded by, and the social or ethical commitments one had.

In telling this story, or stories, I often operate at a much closer range than is customary in a work of intellectual history. I do so in an effort to bring out the perspectives of those involved—the discoveries they thought they were making, the antagonists they thought they were engaging, the stakes they thought they were playing for, the results they thought they had achieved, and most important, the ends they actually served whether they knew it or not. I hope by this means to humanize the process by which this concept, which has figured in the dehumanization of so many, managed to flourish in societies that regarded themselves as advanced, enlightened, and humane. Without such a perspective, the history of the concept of race can seem a strange chronicle of ignorance and bigotry, a kind of zombie that must be killed again and again, rather than what I believe it to be, a story involving people who are intellectually and morally recognizable to us. Those skeletons are in *our* closet.

We are, I believe, dangerously wrong to assume that we have nothing to learn from the history of the concept and nothing to gain by revisiting it. There is a great deal to learn and to gain, especially if we begin with the fact that many of those involved were among the most distinguished scientists and intellectuals of the age. If, as so often happens in recent years, we discount the entire project as nothing more than morally corrupt pseudoscience, we pay science and modernity an undeserved compliment by pretending that honorable people doing real science would have managed the inquiry more responsibly, and that real science could never be tainted by extrascientific ends or interests. The lesson here is rather that the most admirable or impressive minds in any era, our own included, can be tragically blind on issues of great importance. But a further lesson concerns the ability of science to self-correct. When at the end of the nineteenth century many respected figures argued that race was not, after all, a legitimate object for scientific inquiry, they were not condemning pseudoscience in the name of real science; they were simply announcing a result.

The history tracked here begins a century earlier with the dawning recognition that under the rubric of race, human variation could be conceived as a rational system that could, like other phenomena in nature, be studied by empirical methods. When considered as part of the system of nature, the old, rarely-used, and ambiguous term race was transformed from a simple noun with many synonyms into a problem of increasing scope and urgency. This transformation is the subject of the first chapter, which is followed by three chapters, each focusing on a particular episode in which a fundamental question about humanity was engaged through a scientific controversy centering on race.

The second chapter takes up the way in which an emerging science of human variation raised the question of whether humankind should be considered a single group with variations or several groups each with its own distinctive features. A man of the Enlightenment who had corresponded with Immanuel Kant, Blumenbach, who is often credited with the founding of a science of race, was entirely committed to the first position, while his largely despised Göttingen contemporary and colleague Meiners was equally convinced of the second. The proximity of these two men, polar opposites in many ways, enables us not merely to contrast their radically different views but also to explore a deeper and uncanny affinity between them marked by the striking fact that in recent years it has been the enlightened man of science Blumenbach and not the bigoted misanthrope Meiners who has been accused of "racism."

Chapter 3 focuses on the "American School" of anthropology led by Nott and including Samuel George Morton, George Gliddon, Louis Agassiz, and others. The overriding concern of this group in the years before the Civil War was to translate arguments of the kind made by Meiners about the radical differences between Negroes and Caucasians into biological terms by insisting that the two groups had to be considered not as variations within a unified species but as distinct species, a taxonomic refinement that had far-reaching political and moral implications. Pursuing a suggestion made by Blumenbach, the American School placed skeletons and especially skulls at the center of the inquiry into race. The fact that the American School and the practice of craniology were taken quite seriously by the European scientific community provoked a ferocious response from Frederick Douglass in "The Claims of the Negro, Ethnologically Considered," an 1854 address in which he indicted not just Nott, Morton, and others but contemporary science as a whole.

Chapter 4 is centered in Paris, where the question of the equality or inequality of the races came fully into focus. The central figure is the distinguished anatomist Paul Broca, who, having been deeply impressed by the work of Morton and Nott, applied himself to the study of craniology, inventing new angles of the skull to measure and new instruments with which to measure them in the belief that he could by this means identify natural divisions within the category of the human. This approach had the powerful sponsorship of the Paris Anthropological Society, which Broca had founded. Several years after Broca's death, that organization admitted to membership a young Haitian visitor who was to become Broca's most uncompromising antagonist. Appalled by what was passing for advanced thinking in Paris, Anténor Firmin researched and wrote in just ten months a formidable riposte, *De l'égalité des races humaines* (1885), intended as a refutation of Count Joseph Arthur de Gobineau's *De L'Inégalité des Races Humaines* and as a challenge to his fellow members of the Anthropological Society.

Firmin never became part of any mainstream, but as the century approached its end, intellectuals in Paris and elsewhere were coming to the realization that the immense body of statistical information yielded by the measurement of many millions of skulls was not adding up to concrete knowledge of race or indeed of anything except the capacity of science to produce and record data. By the end of the nineteenth century, a professionalized science sought to vindicate itself not by producing positive knowledge but by admitting that in the case of race no such knowledge could be produced, and that race was like phlogiston, a name without a thing, a reference without a referent. Recognizing that the project was proving itself useless as a means to any worthy end, a number of leading figures argued that the term and the subject be dropped from the scientific agenda.

These scientists did not, however, have the last word. The young American scholar W.E.B. Du Bois feared that the interests of black people would be placed at risk if the concept were to disappear from cultural discourse. As a way of rescuing and resuscitating the concept, he proposed a nonscientific definition of race as a composite of physical, cultural, historical, and psychological elements. It was this concept that, Du Bois insisted, had to be "conserved"—not forever, but until such time as "the full, complete Negro message of the whole Negro race" had been delivered.

Du Bois's arguments have been criticized as untenable and incoherent, but his insistence on the advantages to be gained from a robust concept of race has gained many adherents, especially in the United States, where a substantially but ambiguously dematerialized concept of race, invested with a new set of ends by people unconvinced of its existence as a fact of nature, has flourished with the support of intellectuals, theorists, and activists. Whether this concept represents a final and decisive reckoning or simply a turn of the screw remains to be seen.

THE ENDS OF RACE

CHAPTER ONE

How Race Became a Problem

CONSIDER THESE STATEMENTS:

> *"Some have thought fit, to employ the term of races for four or five divisions . . . but I see no reason for this appellation." (Johann Gottfried von Herder,* Outlines of a Philosophy of the History of Man, *1784)*

> *"A man must be blind not to be struck by similitudes between some of the lower races of mankind, viewed as connecting links in the animal kingdom." (Josiah C. Nott,* Types of Mankind, *1854)*

> *"The notion of race . . . is only a subjective notion, a conception within our minds. . . . One questions whether it would not be preferable to be less free with the word race." (Paul Topinard, "On 'Race' in Anthropology," 1892)*

In sequence, these passages mark the beginning, middle, and end of a story about the rise and fall of the concept of race as a subject for scientific inquiry between the late eighteenth and late nineteenth centuries. This book is an attempt to understand how and why this concept emerged, what the stakes of the discussion were thought to be, what resistances it encountered and eventually conquered, what ends it served, and why the most responsible and respected intellectuals and scientists eventually gave up on it. Since this renunciation created an opening for the call by others to preserve and promote a nonscientific concept of race—a call that is still being heeded—it is also intended as a contribution to more recent debates in which race has figured in discussions of history, morality, economics, social justice, and human rights.

From the perspective of the twenty-first century, the value of studying the history of the concept of race is not self-evident. So strongly has the term been linked with specific social practices, legal arrangements, and ideological formations that the concept can easily seem to have no independent history at all, or none worth considering. And so inescapable are the associations of race with colonial exploitation, enslavement, and violence that it might seem almost morally irresponsible to suggest that the term ever had a neutral or uncertain meaning, or that it was ever unimplicated in practices now considered shameful and beliefs now proven to be false.

For understandable reasons, historical discussions of race often produce the impression that the word always meant what it has come to mean, and that the associations that now burden the subject were always in full view. Looking at the past from our present vantage point, we treat statements made long ago by people who had at best fragmentary information about people living far away in isolated cultures as if they were considered reflections on the descendants of those people. In this way, historical figures are held accountable to present standards as if they were speaking directly to and about our world rather than to and about the world they lived in or imagined they lived in.[1]

From our perspective, the most remarkable and disquieting fact about the emergence of the concept of race is that the word had, in the early days, no definite meaning and even no clear rationale for its use. It slipped almost unnoticed into the vocabulary of eighteenth-century intellectual discourse as one term among many for human variation, its entry facilitated by its inoffensive vagueness. There were other terms for human groupings—varieties, tribes, nations, peoples, and stocks, to list just a few of those in English—and other ways of thinking about human variation than as categories in the organizational scheme of nature. The case for race was not self-evident or compelling; Herder was far from the only thinker to be skeptical about whether it was needed at all.

To a scientific or philosophical culture that valued precision, the term may even have seemed a step backward, for it was applied indiscriminately to humans, animals, professions, nations, and plants; even worse, it combined and blurred two meanings that other terms generally separated: a biological notion of genealogy or descent, and a distinct culture or way of life. The Australian ethnologist Bronwen Douglas describes the slurry of meanings associated with race in the early days:

> According to the *Oxford English Dictionary* . . . the etymology of the English term "race" and its European cognates is "uncertain and disputed." The OED derives race ultimately from Italian *razza*, via French *race*, and the semantic history of the English term is entangled with continental

> meanings. . . . The *OED*'s earliest citations date from the sixteenth century when, with reference to man, the concrete noun race signified a family, a kindred, or the posterity of a common ancestor, as in the "race & stocke of Abraham" (1570). More generally, it meant a "tribe, nation, or people regarded as of common stock," as in "the Englishe race" (1572), or served as a synecdoche for humanity, as in "the humane race" (1580).
>
> The primary connotations of consanguinity and shared origin or descent are patent in the several translations of *une race*, "a race," in an early French-Latin dictionary—they include *familia* ("house," "family"); *gentilitas* ("kindred"); genus ("birth"); *sanguis* ("blood," "descendant"); and *stirps* ("stock," "stem," "root," "offspring"). The first French dictionary (Nicot 1606: 533–4) explains that race "signifies origin [*extraction*]," as in "man, horse, dog, and other animal of good or bad race" or "a noble race and house." The *OED* cites parallel English usages from half a century earlier. This semantic conflation of a race with family breeding served to fortify the prerogatives of nobility over populace. The first edition of *Le dictionnaire de l'Académie françoise* (1694, II:364) defines race as "progeny [*lignée*], lineage [*lignage*], origin, all those who come from a single [noble] family." . . . But race/race were minor words in French and English before the late eighteenth century while their German equivalent *Racen* or *Rasse* was a recent borrowing from French and rarely used.
>
> Importantly, however, the sixth edition of the *Dictionnaire de l'Académie* (1835,II:553) also gives an extended signified for race: "a multitude of men who originate from the same country, and resemble each other by facial traits, by external form. *The Caucasian race. The Mongol race. The Malay race.*" The *OED* likewise cites late eighteenth-century and subsequent uses of race to mean "any of the major groupings of mankind, having in common distinct physical features or having a similar ethnic background." These emergent meanings are lexical confirmation of a series of important shifts in the linguistic and ideological significance of race in western Europe from the mid-eighteenth century as naturalists appropriated the term to serve novel taxonomic ends.[2]

In essence, what Douglas describes is the gradual but inexorable reduction and clarification of a once-ambiguous and marginally important term to the point where it possessed a single dominant meaning, referring to a physically distinctive category of humans.

Since the history of a word can shadow the meanings it has at any given moment, it is possible that an older application of the term to animals as well as humans was retained as an unvoiced implication that the entire natural world including humanity was organized in categories of this kind. Such an implication would actually recommend the term to the intellectual culture of the mid-eighteenth century as a way of meeting what Douglas refers to as its "novel taxonomic ends."

Douglass is referring here to the fact that the intellectual culture of the time was beginning to understand the world and especially the natural world from a rationalist perspective as an immense system of categories that could best be approached not through theology, mythology, or philosophy but through empirical research and theoretical reflection. The resulting understanding of the natural order could then be applied to the traditional philosophical project of understanding the human condition. As this possibility came more clearly into focus, the term race began to appear in inquiries into the origin, nature, and implications of human variation.

In recent years the idea that the culture of the eighteenth century produced any significant change in the understanding of human variation has become controversial, with a number of scholars insisting that earlier periods also exhibited a "racial" sense of difference. In their introduction to a volume of readings, *Race in Early Modern England*, for example, Ania Loomba and Jonathan Burton argue on the basis of scores of descriptions of various "others" during this period that even without modern biology or zoology, the early modern construction of difference, applied to Negroes, Muslims, Irish, Jews, and others, was effectively a notion of race, lacking only the word. If early modern accounts of difference stressed cultural factors and later accounts stressed biology, they say, both were describing uncrossable barriers between "us" and "them," and in that respect each can be considered a version of the same thing.[3]

This argument is suggestive, but I believe that it overstates continuities and understates the innovation represented by Enlightenment philosophy and by the empirical methods of modern science.

The replacement of theology and traditional concepts such as the "great chain of being" by biology, ethnology, and zoology as master discourses for understanding the human condition does not represent a minor tweaking. And as the readings in Loomba and Burton's volume themselves suggest, it was not only the word that was lacking in the early modern era. Also absent were other entailments peculiar to race, including the linked premises that humanity could be divided into fundamental groups defined by unalterable traits and tendencies not shared with other groups; that some of these traits and tendencies beginning with appearance but including aptitudes, capacities, and dispositions were transmitted by inheritance and not by culture or tradition; and, most important, that groups themselves were not merely geography- or climate-induced variations on a common pattern but were part of a natural order that many still believed was established by God. Jews and Muslims can convert; races cannot.

In one of the strongest statements of the discontinuity thesis, Ivan Hannaford, whose formidable book *Race: The History of an Idea in the West* begins with a nearly two-hundred-page exploration of "Western History and Thought before Race," argues that race represented not just a new vocabulary of difference but a fundamentally new way of conceiving of generation, descent, and ordering that became, over the course of the nineteenth century, a scientific obsession, encroaching on and eventually displacing politics. What Hannaford demonstrates in compelling detail is that while, for example, ancient Greeks and Romans noted differences in appearance and behavior between themselves and others, they did so without antipathy or stereotyping. The Greeks' sense of their superiority to the Ethiopians, for example, "had nothing to do with the phenomenon of race" and was based rather on the "Homeric ideal of excellence," which the Greeks honored and the Ethiopians did not. To those who would argue for continuity and gradualism in the European understanding of difference, Hannaford responds that "the idea of race does not lie dormant in every society on all occasions and at all times, simply waiting to be discovered. . . . [R]ace is *not* everywhere." Race, he concludes, is an Enlightenment notion, "an attractive and powerful modern idea" that was "inimical to Western civilization in the strict political sense of that word as it was understood before the Reformation."[4]

Race, in short, is not mere difference; it is difference, with a difference. We could say that while the *perception* of difference is a constant in human history, race represents a new and distinctive *understanding* of difference. The early modern accounts of distant peoples, tribes, or nations were *descriptive*, taking the form of lists or inventories of physical features, cultural practices, beliefs, languages, locations, modes of living, deities, diets, and so forth. Many of the studies of human variation conducted in the later eighteenth and early nineteenth centuries under the rubric of race were, by contrast, *analytical*, and took the form of an inquiry into the origin, depth, nature, and consequences of the physical and other differences among humans. The study of races focused on the possibility of a system of such differences, a hidden organizational scheme of nature that when fully articulated would have philosophical, social, political, and even theological implications. The concept of race reflected a new understanding of human variation not just as something to be recognized and described but as a problem—an important problem with tremendous implications—to be solved.

In recent years, many have argued that the concept of race was itself a problem, not a solution. The general accusation is that by promoting the idea that racial categories were facts of nature, nineteenth-century science wittingly or unwittingly consented to a fantasy of fixed biological categories that served the ends of colonial domination and economic exploitation. The entire inquiry into race, according to this argument, was from the outset a corrupt enterprise undertaken for political or ideological purposes, a pretentious representation of difference functioning as a pretext for the imposition of power by the more advanced and *soi-disant* enlightened nations over the benighted, backward, mentally limited, and spiritually deficient populations of the world, especially those in what is now known as the Global South.

In 2019, the American Association of Biological Anthropologists made this argument against its own discipline, declaring that "the Western concept of race must be understood as a classification system that emerged from, and in support of, European colonialism, oppression, and discrimination. It thus does not have its roots in biological reality, but in policies of discrimination." In a recent book, Joseph L. Graves Jr. and Alan H. Goodman boil this case down to its essence, describing the concept of race bluntly as "a myth that provides cover for racism." The prominence both early and late of German intellectuals in the discourse on race strikes some as particularly telling evidence of the complicitous implication of the concept in atrocities. Introducing a collection of nineteenth-century writings, Hannah Franziska Augstein defines nineteenth-century racial theory as "the biologically founded racial theory . . . culminating in the racial ideology of the German National Socialists." Whatever the role their countrymen may have played in the invention of race, however, German thinkers are now taking the lead in denouncing the entire project. As the 2019 "Jena Declaration" issued by the German Zoological Society proclaims, "The concept of race is the result of racism, not its prerequisite."[5]

This charge is certainly not unsupported. Nobody could dispute that various theories of hardwired differences among peoples were instrumental in justifying programs of systematic disenfranchisement, domination, and violence carried out by some Western powers against populations that could be represented as being so different from them that they neither possessed nor deserved status or rights comparable to their own. It is also indisputable that by the mid-nineteenth century in Europe and the United States, the concept of race was figuring in debates around slavery, the imperial conquest of indigenous peoples, and political revolution. Especially from the perspective of the twenty-first century, it is easy to see how some discourses on race made explicit or implicit accommodations with the premise of inequality used to justify these programs, and possible to understand that some would regard the defense of that premise as the point of the entire undertaking.

But attitudes, beliefs, and practices we now call racist including slavery have been a constant feature of human existence, and have never required philosophical or scientific support, much less a "myth that provides cover" for them. American slavery was an extraordinary stimulant to racial theorizing, but the New World slave trade flourished for two centuries without a theory of race. And while many intellectuals who contributed to a theory of racial difference—Edward Long, Julien-Joseph Virey, James Hunt, Christoph Meiners, Robert Knox, Thomas Henry Huxley, Samuel George Morton, William Lawrence, Josiah C. Nott, Ernst Haeckel, and Paul Broca—were convinced of both the reality and the inequality of races, and of the inferiority of black people in particular, and while as Frederick Douglass noted in 1854 some of their work was cited by supporters of American slavery, the institution of slavery received little explicit support from respected intellectuals outside the United States during the antebellum period.[6] Even the proslavery advocates Hunt and Nott could only endorse the practice on the humanitarian grounds that slaves were better off in the United States than they had been in Africa where many of them had already been enslaved or subjugated before they were sold to slave traders, or the economic grounds that slavery was essential to the prosperity of the nation, or the practical grounds that slavery was the only workable relationship between such different populations. Just as common as tacit complicity with slavery or other forms of violence—and far more common than explicit assent—were statements of revulsion or repudiation.

Not even those who felt that the differences between races were effectively fixed necessarily believed that such differences justified acts of violence, domination, or expropriation, of which they were acutely aware. Some felt that present inequalities could be eliminated over time given equal conditions and opportunities; others believed that inequality was an argument against rather than for domination. Most if certainly not all of those whose contributions on the topic were taken seriously by the intellectual community rejected on moral grounds any attempt to use their work to support the economic or imperial domination of one racial group over others.

That group included Charles Darwin, a man of broadly humanitarian convictions whose *On the Origin of Species* (1859) could be said to have lent itself to projects of exploitation or colonial domination by positing a process of gradual modification from lower or primordial forms to higher, more specialized and complex forms. One could conclude from Darwin that some populations were more primitive than others and were for that reason either fitted for exploitation or doomed to extinction. But this was not a conclusion Darwin ever drew or came close to drawing; nor did he endorse the assimilation of race to species on which many such arguments rested, stating in *The Descent of Man* (1871) that the fact that groups "graduate into each other" is "the most weighty of all the arguments against treating the races of man as distinct species."[7]

Darwin was undoubtedly thinking of the arguments coming out of the "American School" of anthropology led by Nott, which will be discussed later in this book. But others who agreed with Nott on the question of a fixed Negro inferiority also rejected such arguments. According to Thomas F. Gossett, the majority of proslavery advocates in the antebellum South believed that a species distinction between blacks and whites contradicted the Bible's story of the origins of humanity, and given the choice between the Bible and a science that might have provided the only intellectually respectable defense of slavery they could have taken up, they chose the Bible and accepted the humanity, if not the equality, of slaves.[8]

Even if the scientific community had the will or desire to support policies of racial domination or exploitation, it would have been an ineffective ally. This at least is the claim of historian Emmanuelle Sibeud in her study of the Anthropology Society of Paris founded by Broca in 1859. This group, which was in its early years dedicated to the study of skulls in which Broca had become a leading figure, is perhaps the lowest-hanging fruit for the argument that science was complicit with racism, but as Sibeud shows, the Society was by the 1880s already in a state of disarray as a consequence of Broca's death and the internal wrangling that followed. Burdened by the manifest failure of craniology to yield convincing evidence of anything, and still unable to define race itself, the Society was "neither an efficient colonial tool nor a racist plot projected from the metropolis on the colonies." All in all, it was, in the words of her title, a "useless colonial science."[9]

Nor, according to Colin Kidd, did European powers ever require scholarly or scientific approval for their exploitation of non-European peoples. When Europeans undertook the exploration and conquest of much of the rest of the world, they did so, he argues, out of a mixture of private and national motives, and relied for conceptual support and warrant not on any sense of biological differences among populations but rather on what Kidd describes as "the Christian imperative to spread the Word of God through missions." If validation were required, it was sought not in racial science but in "theories of natural law concerning the connection between proper use of the land and rights to its ownership. . . . 'Whites,'" Kidd concludes, "conquered the world without any overt ideology of white superiority," and indeed without any understanding of themselves as "white."[10]

And so we return to Bronwen Douglas's assertion that race was a response not to the ideological interests of colonialism or the slave trade but rather to the scientific culture of the eighteenth century in which we recognize the origins of intellectual modernity. The most compelling instance of the categorical ordering of nature and one of the signal achievements of the modern ethos Hannaford refers to is Linnaeus's *A General System of Nature*. Appearing in twelve editions during the author's lifetime with the definitive tenth edition appearing in 1758, this treatise provides a dazzling, if speculative and continuously evolving, model for understanding the complexity of the natural world.

It is surely Linnaeus that Douglas is thinking of when she mentions the "novel taxonomic ends" of the intellectual culture of the eighteenth century. By including humanity in the "general system" of nature, Linnaeus placed human beings in a logical order with classes and subclasses that could be determined by the close observation of particulars coupled with an imaginative grasp of the total system. Race—a concept unavailable to Linnaeus—would become an important part of that system as others tried to improve on his astonishing but idiosyncratic and preliminary efforts.

Although Linnaeus wrote in Latin well before the use of the term race had become commonplace, he—or rather his English translator—had inadvertently established the framework of the debates about human variation that would unfurl over the next century and a half. Linnaeus's full title mentioned the three "grand kingdoms of animals, vegetables, and minerals," divided into what was called, in Latin, their "*Classes, Ordines, Genera, & Species*"; his translator William Turton added "varieties," creating at a stroke a small pocket of uncertainty and hence opportunity for later thinkers who would try to slot race into the system of nature.[11]

Lacking a precise definition, the term seemed to oscillate between species and varieties, its neighbors in the Linnean, or Turtonian, system. When, in *Notes on the State of Virginia* (1787), Thomas Jefferson wrote, "It is not against experience to suppose, that different species of the same genus, or varieties of the same species, may possess different qualifications," he anticipated the terms of the following century's debate on race, which centered on the question of whether race was more like a species

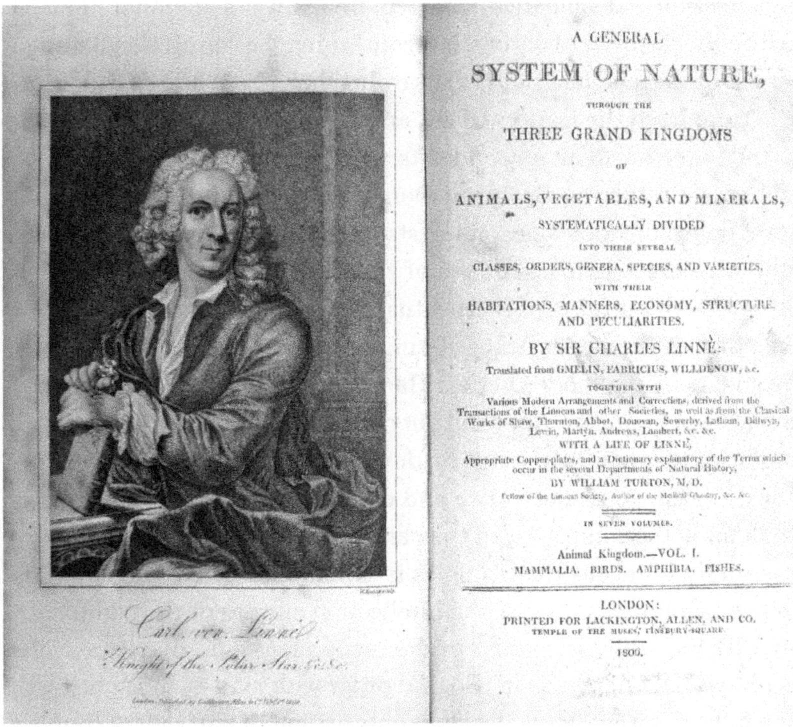

FIGURE 1.1. Sir Charles Linné (Linnaeus), *A General System of Nature*, ed. and trans. William Turton (1806).

or more like a variety.[12] This ambivalence proved to be of immense consequence. Over the next century and a half, the discourse on race hovered over this question: Was *Homo sapiens* a species with races constituting environmentally induced varieties, or was it a genus, with races constituting stable species? Other questions followed in the wake of this one: What is the relation between the adjacent genuses of *Homo* and *Simia*? Were some races closer in some significant sense to *Simia* than others? And how were categories organized? Were they just differences, or did they have a range, with a top, middle, and bottom?

Linnaeus's successors were aided in their efforts to puzzle out the meaning of race by an accumulating mass of information about distant populations that was pouring into Europe from travelers, missionaries, pilgrims, maritime merchants, explorers, and commercial or colonial officials. For such people, differences were facts to be recognized and negotiated in order to be able to trade information and goods with the locals. But for European intellectuals, the information they reported was considered as evidence that could be used to assemble a comprehensive scientific taxonomy.[13] The reports themselves tended to dwell not on categories but on exotica, leaving to the philosopher or scientist (the distinction not always sharply drawn) the task of extracting regularities and patterns from often exaggerated tales of nature's profusion. The construction of a full and accurate account of the place of humanity within the natural order and the articulation of the distinctions within the human species represented not only a challenge for science but also—given Linnaeus's binomial definition of humanity as *Homo sapiens*—a species imperative, a consequence of the injunction to Know Thyself.

The ambivalence of the term gave race a flexibility that enabled it to survive challenges even as it limited its effectiveness as a descriptor. When, for example, it was objected that the peoples of the world blended into each other with no bright lines dividing one group from another, the concept of race was able to survive by becoming a system of types, ideal essences that did not imply discrete and identifiable human populations. This concept, announced in unmistakable terms by Nott and George R. Gliddon's 1854 compendium *Types of Mankind*, seemed to promise a certain kind of clarity, but in time it became apparent that the innovation was itself unclear because while it appeared to suggest a norm from which deviations could be marked, the norm was difficult to locate.[14] Debates about race became debates about the site of the norm, whether it was in the distinctive qualities of a group, the average member of the group, or the ancestors of the group. None of these had any authoritative claim, however, and none could in any event be determined with any precision.

The typological conception of race may not have settled the many issues surrounding the subject, but it did generate a remarkably durable feature of racial science that has come to seem particularly sinister, a preoccupation with quantification and classification, as millions of people were studied and measured in various ways over the course of the nineteenth century and beyond in an effort to develop the data set for racial types. The reductive implications of these procedures were not at first generally recognized; what was apparent was the dramatic increase in conceptual power enabled by the emphasis on the empirical, and the promise at least of the categorical. Over the course of the nineteenth century, the gathering, observing, measuring, and classifying of human bodies and human populations would become a routine aspect of medical research whose statistical and objective character bespoke a scientific rigor.

Many such efforts today seem outlandish, like British scientist John Beddoe's attempts at the end of the nineteenth century to devise an "index of nigrescence," an algebraic equation measuring the precise degree of darkness of the skin of inhabitants of the British Isles (Celts scored higher than Anglo-Saxons), or William Z. Ripley's discourse in *The Races of Europe* (1899) on the "nostrility" characteristic of Jews.[15] But they did not seem so at the time, and the combined if uncoordinated efforts of various methods and approaches had the cumulative effect of creating the concept of race as a highly plausible and even imposingly rational way of thinking about the differences among the peoples of the world.

The dominant approach, and the one to which the most attention is paid in this book, was craniology, or the measurement of skulls. At once the most dubious by current standards and the most sinister of the methods used to pursue the subject of human variation, craniology had advantages that other approaches lacked. Skin pigmentation was felt to be too variable and graded to yield useful information—it was, as Gossett remarks, "one of the least satisfactory of the methods tried" to determine racial categories—and brains were too subject to degradation.[16] Skeletons had marrow, which could also degrade, and tended to fall apart. Skulls, by contrast, could be easily gathered, stored, cataloged, transported, and studied at leisure. Providing the hardest of hard evidence, skulls were an exemplary object of a scientific inquiry that had extrascientific implications and rewards for the ambitious scientist.

Ambition was required for such a vast undertaking. Travelers were often urged by anthropologists to bring back measurements that could be added to the archive. The director of the Anthropological Museum in Budapest near the end of the nineteenth century, Aurel Török, is reported to have advocated five thousand different measurements for every skull, some to be made with an invention of his own devising, the gnathometer. For many physicians, each patient was, in addition to whatever else he or she might be, a skull. Before he sets out on his voyage into the heart of darkness, Joseph Conrad's narrator Marlow has his skull measured "as a simple formality" by a doctor who "produced a thing like calipers and got the dimensions back and front and every way, taking notes carefully," saying, "I always ask leave, in the interests of science, to measure the crania of those going out there." He had, the doctor tells Marlow, "a little theory which you Messiers who go out there must help me prove." Craniology was part of both the imperial and the scientific projects; as the doctor confesses, "This is my share in the advantages my country shall reap from the possession of such a magnificent dependency."[17]

One cannot read many of the early treatises on race—even the most deeply flawed or reprehensible—without sensing the excitement the authors felt at what seemed a magnificent opportunity to establish the ability of empirical research and scientific method to engage with issues whose interest and importance could be immediately grasped by the public and even valued by the state. Those engaged in the study of race experienced the exhilarating challenge of developing a new field of research, the satisfaction of contributing to a growing archive of fact, and the privilege of making meaningful scientific contributions to questions of both immediate and enduring interest. Phrases such as "the ladder of existence," "the great chain of being," and "the order of creation" had been part of the European vocabulary for centuries.[18] The study of race, sometimes framed as a scientific inquiry into the "scale of nature," promised to yield the hidden logic or principle of order—the laws of nature—that undergirded these phrases, making the old concepts newly meaningful to a modern culture. As a term that suggested science and modernity but whose implicative reach exceeded its referential grasp, race could open up new approaches to a range of important religious, philosophical, and scientific issues that were beginning to be framed anew in a culture of intellectual modernity: the universality of human nature, human understanding, human reason, or human rights; the limits of species plasticity; and the place of mankind in the order of nature.

The primary issues to be explored were the origin, character, and depth of human differences, but useful work could also be done on the specific roles of geography, climate, diet, language, and culture in creating, consolidating, or aggravating those differences. If light could be shed on these questions, then the even more mysterious purpose of difference, a question heretofore relegated largely to theology, might be resituated within the domain of science. Then, if the enigma of race could be solved, it might be possible to determine whether all humans possessed the same or different capacities, and thus whether the differences among groups represented alternative realizations of a common human potentiality or the full expression of different potentialities. It would become possible to say once and for all whether humanity was one, as both the Church and classical cosmogony taught, or many, as learned opinion increasingly held. Light would be shed on the difficult question of the responsibilities of the more advanced and powerful cultures toward the peoples they were exploring, colonizing, exploiting, or enslaving.

The urgent questions surrounding the issue of how to think about these peoples—and necessarily, how to think about one's differences from, connections to, or obligations toward them—required a vocabulary of difference. Race, an uncertain concept of relatively recent vintage, had emerged from a crowded field as the strongest candidate for the study of human variation because of its potentially comprehensive explanatory power. Unlike any other term, race combined, or seemed like it might combine, elements of biology, culture, geography, and nature in a single and singularly powerful concept.

As consequential as the stakes were, however, the inquiry into race did not lead in only one direction; indeed, the various meanings of this conflicted concept disclosed fissures within the Enlightenment ethos itself. As what Bronwen Douglas called "a synecdoche for humanity," race was entirely consistent with the universalizing thrust of Enlightenment philosophy; as a system of subcategories within humanity, however, race was an artifact of science. The second meaning negated the first because it accorded any subgroup that could be described as a race the same status as the larger group, so that "Caucasians" could—and frequently did—say they were as distinct from "Ethiopians" as humans from primates. It is the scientific tendency to proliferate distinctions that Hannah Arendt was thinking of when she identified race with "the exclusion in principle of the idea of humanity. . . . Race," she said, "is, politically speaking, not the beginning of humanity but its end."[19]

To make matters worse, the invented and therefore arbitrary character of all the terms and concepts in play was acknowledged from the beginning, or even before the beginning. In 1749, before any serious effort to explore the concept of race, the French naturalist Comte de Buffon allowed that categories such as type, variety, stock, and family were more or less interchangeable, and were in any event things "of our own creation. . . . [W]e have contrived them to ease our memories, and to aid our imagination."[20] The American naturalist Samuel George Morton began his *Crania Americana* (1839) with the acknowledgment that "the grouping of mankind into Races, has occupied the ingenuity of many of the best naturalists of the past and present century" without producing any consensus even on the number of races, with proposals ranging from one (the human) to fifteen.[21] Morton himself accepted Johann Friedrich Blumenbach's identification of five basic or primordial races, but identified within them twenty-two subordinate "families"—noting, however, that families were often confused with races, and that some scholars used the word race when they meant variety.

Morton's successor in some respects, Nott complained about the difficulty of making any argument about the natural system when the terms themselves were so obviously made up. "The embarrassment in treating of types or races," he said, "is constantly increased by false classifications." Each term or principle of classification had resulted in a new number:

six varieties (Buffon), eleven races (Charles Pickering), sixteen species (Charles des Moulins), and sixty-three races (Luke Burke, who included in his total twenty-eight "intellectual" and thirty-five "physical" races). Like many others, Nott himself was unembarrassed by his own inventions, assuring his readers that his identifications of races and types and types with species represented "facts or laws which have emanated from the Creator."[22]

Nott was drawing on a well established understanding of species that could trace its pedigree to the seventeenth-century English naturalist John Ray, who defined a species as an unchanging product of an original creation. What seems to have escaped Nott's attention was the fact that this definition of species as unchanging types had come under stress, as studies of crossbreeding and hybridism were beginning to suggest the theologically scandalous possibility that species were not fixed and unalterable but might become transformed over time, and even that new species might have come into being since the creation of man. With this possibility in the air, the concept of species had become exposed to the accusation that it, like race, had been imagined into being not by God but by human beings, and could be imagined differently.

We can actually observe the moment when the concept of species defined as a clear and distinct category begins to blur or melt when it comes into contact with race. In the 1826 edition of his *Researches into the Physical History of Mankind*, the distinguished British naturalist James Cowles Prichard begins a meditation on whether all humans belong to a single species with a preliminary comment on the term species, which he says is "very simple and obvious. It includes only one circumstance, namely an original distinctness and constant transmission of any character. A race of animals, or plants . . . which have always been constant and undeviating, constitutes a species."[23] Simple and obvious enough—except that Prichard conflates species and races, and then compounds the confusion in the very next sentence by noting that "the same meaning was originally attached to the term genus, which we now apply to species." Animals themselves, he concedes, seem confused about what they are, because while the fundamental criterion for any classification is resemblance, oxen, bison, and buffalo "bear a striking resemblance to each other," and some variations that have become heritable have caused species to differ from themselves so that they resemble other species. In short, "though the idea of species is simple and definite, it is not always easy to determine what races of animals are of one, and what of distinct species."[24]

We know there are species, there must be species—the idea is so beautifully clear—but the animals themselves were making a barnyard mess of the whole thing.

The consequences of terminological ambiguity and inconsistency were especially pronounced in the United States, which was, as its founding documents proclaimed, philosophically and politically committed to an Enlightenment-based concept of human oneness and equality, but which, as the institution of slavery demonstrated, was also committed to the principle of radical differences and permanent inequalities among groups. If races were considered to be stocks, families, tribes, or varieties within a single species, all sharing a common ancestor, Enlightenment premises were not threatened, but if races were more like species in a larger genus, then perhaps some creatures who, as Prichard says, "bear a striking resemblance" to humans might not actually be humans. They might be another species altogether—which would explain why some white humans particularly in the American South might feel that they were living alongside strange and mysterious beings with whom they shared no ancestry and had nothing in common other than that they were diurnal, walked on two feet, and had lactiferous teats. In that case, a different philosophy and even a different politics and a different morality would have to be developed to deal with these humanlike creatures, who had at best a very restricted access to human reason, human understanding, or human rights.

Accommodating both implications, the concept of race—referring both to all of humanity and to groups within humanity, and wobbling between species and variety—registered this ambivalence without resolving it, and so could be deployed in arguments on either side of the question.

A science of racial difference presented a different kind of challenge to Christian theology. By inquiring into the origin of human differences, students of race were trespassing on ground long claimed by the Bible, whose commanding account of human origins still functioned for many as gospel truth. At first, the possibility of a rivalry between science and religion was generally denied, as many of those who wrote about race—John Mitchell, Petrus Camper, Sir William Jones, Samuel George Morton—labored to reconcile their findings to the terms of established religion, with existing human populations being credited with a genealogy that extended back to Eden or, if that seemed implausible, adduced as evidence of the inscrutability of God's intentions or mysteriousness of God's ways.[25] Debates about race in the early nineteenth century often took the form of clashing interpretations of Genesis. The "monogenists" who held that the human species was essentially unitary could point to the single act of creation and a single primal pair, while "polygenists" who believed in separate creations and therefore in a deep division between groups could argue that the Bible described only the experience of one tribe among others whose existence is suggested by, for example, the unexplained appearance of the wife of Cain or the wives of the sons of Noah.[26]

As mid-century approached, however, some began to feel that serious thinking about race required either a concession that the Bible was an ancient fable that translated God's truth into human terms, or, more radically, a wholesale rejection of the Biblical story as a myth or a compendium of pious lies. Nott's sneering dismissal in *Types of Mankind* of the notion of a "primal pair" of each species—"a *pair* of bees, locusts, herrings, buffaloes"—suggests a professional exasperation with the constraints on science placed by the conviction that the Bible was a source of historical truth.[27] For Nott and others, race was the point at which religion and science threatened to collide or part ways; for a growing number, it was the point on which religion had to yield to science, which was seen by some advanced thinkers not as a complementary but as a superior way for a modern culture to understand human origins. And so, over the course of the century, a Bible-based monogenism consistent with what Arendt describes as the Enlightenment "idea of humanity" was replaced by a harsher, biology-based polygenism that argued that the various races had always—or at least for an extraordinarily long period of time—been as they were at present.

Even so, the authority of the Bible was so durable that as late as 1943, the anthropologists Ruth Benedict and Gene Weltfish felt they had to reassure readers of their pamphlet "The Races of Mankind" that the result of post-Darwinian scientific research into race was that science and Christian faith were in essential agreement, and that "the Bible story of Adam and Eve, father and mother of the whole human race, told centuries ago the same truth that science has shown today: that all the peoples of the earth are a single family and have a common origin."[28]

Race had a more complicated effect on one of the other central issues in the nineteenth century, the concept of the nation. Both concepts emerged, or were invented, at roughly the same time, in the wake of the American and French Revolutions and, a generation later, the dissolution of the Holy Roman Empire. These events forced people living in the affected areas to understand themselves not as subjects who merely happened to live on land claimed by a powerful sovereign but as citizens who, collectively, comprised the nation. As a potentially powerful new way of thinking of collective identity, race found its way into many debates on this subject.

In its unsettled state, race played more than one role in the various attempts to conceptualize the nation, and was conscripted on behalf of different causes or ends. In a brief but probing account of "Race-Thinking Before Racism," Arendt argues that at the beginning of the nineteenth century, race was frequently aligned with an antimonarchical liberalism, but became, especially after the revolutionary events of 1848, part of a "nation-destroying and humanity-annihilating" ideology whose only rival in this respect was Marxism-Leninism.[29] Each of the Western nations, she says, followed its own version of this trajectory, cultivating aspects of race that served its needs and interests.

Even within a single country, these needs and interests could be complex or contradictory. The understanding of French citizenship prevailing for most of the nineteenth century treated race primarily as an instance of the kind of divisions among peoples that a national identity based on liberal or revolutionary principles could accommodate by overriding them. Jules Michelet began his nineteen-volume *Histoire de France* (1831) by describing his nation as composed of "race upon race, people upon people"; fifty tumultuous years later, Ernest Renan was still celebrating the ability of the French nation to absorb and assimilate groups that might otherwise be considered natural antagonists including Franks, Normans, Bretons, Celts, and Gauls. A nation that was, as Renan put it in his famous 1883 essay "What is a Nation?," "Roman by language, Greek by civilization and Jewish by religion" was defined by its ideals not by its racial composition.[30] France exported these ideals to its colonies including the Arabs in North Africa, the Africans in Africa and the Caribbean, and the Vietnamese and others in Indochina through its language, its culture, and its administration, encouraging people living there to think of themselves as Frenchmen and Frenchwomen despite the fact that few if any of them had Roman, Greek, or Jewish—or French—ancestry.[31] In the French instance, then, the "idea of humanity" that according to Arendt was "excluded in principle" by the concept of race was actually strengthened by acquiring the committed patronage of the state, whose heroic mission was not to deny racial differences but to render them irrelevant.

This spacious view of a French nation in which race counted for less than national identity was, however, not shared by the dispossessed post-revolutionary aristocracy, which registered its despair at the direction of the nation by forging a tribal bond with the aristocracy of other nations, an imagined *Internationale* of inherited nobility constructed on a phantasmatic all-but-racial basis. In the most fully developed instance of this argument, Comte de Gobineau's multivolume tome *The Inequality of the*

Human Races (1853–55), traced all honor, nobility, and virtue back to a long-vanished horde of "Aryans," an all-conquering horde that had roamed the world, mixing its blood with the local races. This mixture had been fatal to the Aryans, who had vanished as a consequence, but it was a genetic boon for the favored locals. In France, Gobineau said, the Frankish nobility represented the direct descendants of this aboriginal aristocracy and were for that reason superior to the lesser breeds, the Gauls and Celts who made up the rest of the French citizenry.[32] For Gobineau and others of his persuasion, the concept of race did indeed function, as Arendt said, as a "weapon for civil war and for splitting the nation."[33]

But race could be weaponized in in the service of other ends as well. The liberal-seeming Renan, with his magisterial view of the history of humanity, was, Arendt says, nevertheless "the first to oppose the 'Semites' to the 'Aryans' in a decisive '*division du genre humain*,'" a division that would become more pronounced at the end of the century.[34] And Renan and others defended French colonialism in Africa and elsewhere by invoking natural differences among peoples that could easily be construed as racial differences. And while some such as Hippolyte Taine argued for a natural alliance between France and Germany on racial grounds, others especially after 1870 insisted on a natural division between the same nations—also on racial grounds.[35] More than in any other country, race became in France a kind of Swiss Army knife, fitted for a wide range of uses.

It was in Germany, however, that the astonishing shape-shifting range and flexibility of the concept was most fully explored. The first appearances of race in German intellectual discourse often had liberal associations of the kind exemplified by Blumenbach, but when conjoined at a later date with other ideas, race took on a darker, more divisive aspect. At first seemingly opposed to nationalistic sentiments, the concept of race was subsequently reworked in a way that established a deep compatibility with the Romantic nationalism exemplified by Herder, who promoted an organic or naturalistic understanding of "the people" (*das Volk*) as an indigenous population united not by subjection to a sovereign but by common origins, language, cultural traditions, and blood relationships. In some of the more mystical articulations of this concept, the people were represented as virtually autochthonous, with a character as ancient and unalterable as the land itself. At the end of the nineteenth century, this principle of national unity was translated into an explicitly racial nationalism that became, in the twentieth century, the "humanity-annihilating" ideology Arendt describes as Nazi "race-thinking."[36]

Having endured wave after wave of foreign invasions since the Romans, Britain could not, at first, look to race as a source of national identity. History not genealogy was understood to constitute the singular character of the British people, whose defiantly mongrel composition was both satirized and celebrated over the course of a thousand exuberant and even manic lines in Daniel Defoe's 1701 satire, *The True-Born Englishman.*

Thus from a mixture of all kinds began
That heterogeneous thing, an Englishman:
In eager rapes, and furious lust begot,
Betwixt a painted Britain and a Scot:
Whose gend'ring off-spring quickly learn'd to bow,
And yoke their heifers to the Roman plough:
From whence a mongrel half-bred race there came,
With neither name, nor nation, speech nor fame.
In whose hot veins new mixtures quickly ran,
Infus'd betwixt a Saxon and a Dane;
While their rank daughters, to their parents just,
Receiv'd all nations with promiscuous lust.
This nauseous brood directly did contain
The well-extracted blood of Englishmen. . . .
We have been Europe's sink, the jakes where she
Voids all her offal outcast progeny.[37]

In Defoe's immensely popular rendering, the English national character was anchored not in a system of lineal nobility in which some were by nature elevated over others but in a regnant populace in which none could legitimately claim to be more "true-born" than another. While other nations might assert their distinctive identities in a way supported by the concept of race, many British citizens thought of themselves as white, Caucasian, Anglo-Saxon, or European, categories they shared with millions of others including some with whom they went to war. British national identity, a produce of evolution not revolution, and always less sharply defined than that of France, was at this time and for many years thereafter grounded not in race but in the monarchy, geography, speech, and politics, as well as in cultural, legal, and intellectual traditions.

Over the course of the nineteenth century, however, this understanding was challenged by various racial theories that connected physical features to cultural behaviors and even to inbuilt capacities and traits. By midcentury, a strain of Anglo-Saxonism had, in the words of George W. Stocking Jr., begun to take on a distinctly racial meaning, resulting in claims of biological as well as historical and cultural connections between Britons and other Germanic or "Teutonic" peoples, and a natural antagonism between the liberty-loving, mature, and self-reliant Anglo-Saxons and the wild, impulsive, dreamy, and violent Celts.[38] The concept of race, and of an "Anglo-Saxon" race in particular, became commonplace, popularized in part by a series of novels by the future prime minister Benjamin Disraeli. A character in Disraeli's 1847 novel *Tancred* insists that "all is race; there is no other truth"; in the same vein, the Scottish anatomist Robert Knox, speaking for himself in 1850, gave voice to an extreme form of racialism conceived as a suite of physical and other characteristics unchanged since the beginning of time, declaring bluntly that "race is everything."[39]

British intellectuals of a more cosmopolitan and humane disposition found reasons and ways to agree with Disraeli and Knox about race as a factor in culture and history. In a country with colonial possessions in East Africa, India, Burma, and the Middle East, the conviction that features fixed in nature determined a people's fate in the world was not out of the mainstream. In an 1870 essay titled "The Forefathers and Forerunners of the English People," Thomas Henry Huxley gave voice to this form of common sense, insisting on "the immense influence of that fixed hereditary transmission which constitutes a race."[40] Supported by the developing practice of craniology, the growing scientific endorsement of polygenism, and the emergence of professional disciplines of ethnology and then physical anthropology, race established itself as a highly plausible way of accounting for many cultural or biological differences.

In no country were controversies surrounding the racial composition of the nation more directly consequential than in the United States, a "nation of immigrants" with a history of race-based slavery and a deeply persistent strain of hostility toward peoples—Jews, Muslims, Asians, Hispanics, Italians, Africans—who could be described as interlopers in an otherwise homogeneous polity. One episode in this ongoing struggle will be explored in chapter 3 of this book, but the larger point of all the debates about race and national identity in the United States and elsewhere concerns the extraordinary adaptability of the race concept, which could be pressed into service in support of a wide variety of ends, some opposed to or in conflict with others. The specific meaning of the concept varied from context to context. As George Frederickson commented in his 2015 *Racism: A Short History*, "Racism is always nationally specific."[41]

In addition to its roles in debates about science, philosophy, religion, and national identity, the concept of race made another less visible kind of contribution to the world we now take for granted: the formation of academic disciplines, specialized ways of knowing with their own assumptions, methods, and goals. Much of the rest of this book will be concerned with the ways race was conceptualized by academic and scientific discourses, but it is clear that the concept of race had a transformative effect on these discourses themselves.

Tracking the changing understandings of the premises, evidence, and methods with which the subject of race was approached, we simultaneously track changes in the understanding of science and the scientific method. Before the mid-nineteenth century, race typically figured as a subordinate element in expansive treatises on such subjects as the natural history of mankind, the philosophy of the history of man, the theory of the earth, human physiology, the physical history of man, or the order of creation. From today's perspective, such sprawling explanations-of-everything, with their charts, graphs, illustrations, and tables, lack the kind of consistent methodology, delimited fields of objects, and clearly specified aims that could produce results that command respect as statements of fact. The disciplines into which modern knowledge has been divided had not yet crystallized or established themselves, so a single text might include materials that would now be distributed across a wide range of fields—archaeology, ethnology, comparative anatomy, zoology, philosophy, theology, geography, dermatology, linguistics, biology, physical or cultural anthropology, physiology, paleontology, phrenology, and craniology—along with witnessed, reported, or fabricated accounts of peoples around the world. In many early discussions, race was one of several terms deployed in a discourse that aspired to comprehensiveness with no sense that boundaries were being transgressed or a necessary precision sacrificed. An instrument rather than an object of analysis, the concept of race was used by people who were not focused on the subject or even necessarily convinced of its existence but found it useful to think about or with.

The problem of race did not begin to attract attention as a separate subject that could be approached by specialized methods until the mid-nineteenth century. We can actually see this trend toward specialization taking root in Morton's 1839 *Crania Americana*, which begins with a ninety-five page introduction, "On the Varieties of the Human Species," that is entirely generalist in its travelogue approach. This is followed by two hundred pages of diagrams and instructions on how to read representations of skulls in a way that would enable one to understand the relations among the parts, and to compare them with skulls from other groups. Following this are nearly ten pages of closely printed measurements, an appendix by a professional phrenologist, and seventy-one detailed lithographs. Phrenology, already moribund, died a quiet if prolonged death over the course of the nineteenth century, and craniology did not survive long into the twentieth, but as we will see, the fields of biological and physical anthropology can trace their origins to these early efforts to understand the large topic of human variation through the systematic study of skulls.

As mid-century approached, societies in ethnology and anthropology were founded in Paris, Berlin, New York, and London, with the Paris society explicitly devoted to craniology.[42] At first, these societies had a philanthropic dimension—the London group, for example, had been incubated in the Aborigines' Protection Society that had been headed by James Cowles Prichard. But these learned societies quickly became professional organizations, sponsoring lectures, debates, meetings, and publications by scholars or scientists including Prichard, William-Frédéric Edwards, Hunt, Knox, Alfred Russel Wallace, Jean-Louis Armand de Quatrefages de Bréau, Carl Vogt, Broca, Paul Topinard, Rudolf Virchow, and Huxley.

After 1859, these societies led the debates about the implications of Darwin's *Origin of Species*, which placed the study of race on an entirely new, predominantly anthropological footing. If race was, as Stocking has written, "the central theoretical concern of pre-Darwinian anthropology," it was also a primary concern of post-Darwinian thinking, which, however, assumed a more modern and disciplinary aspect.[43] While the polymath Ernst Haeckel might deal with race in the course of a massive *History of Creation* (1868), a more restricted and empirical approach is signaled by titles such as *On the Phenomena of Hybridity in the Genus Homo* (1860) by Broca, "Formation of Human Races Under the Sole Influence of Conditions of Life and Heredity" (1877) by Quatrefages, and "Heredity and the Formation of Races" (1896) by Virchow.[44] Such titles announce a segmented approach to knowledge that renounces any grand holistic or synthetic ambitions—the history of creation is, in these works, nowhere in sight—in favor of a research-based approach to a limited and carefully framed set of questions about a single aspect of the whole.

The premise of the discipline is that focused studies on defined problems will contribute their mite to a growing archive of information that could one day constitute a comprehensive understanding that would by design always remain beyond the reach of the discipline. Within a discipline, no work however magisterial, comprehensive, or authoritative is conceived as a final word; there is always more to say, always the possibility of improvement or advance. Even the greatest monuments of modern scholarship are still provisional and preliminary, intended as contributions to a final synthesis whose realization lay in the future. This understanding of the organization and aims of knowledge was ideally suited to the subject of race, which, never having been satisfactorily defined, could not be studied frontally, in itself, but could be broken down into manageable bits—not just nigrescence and nostrility, but skeletal measurements; the fertility of hybrids; the numerous indexes, angles, arches, dimensions, and proportions of the skull; the influence of geography, diet, culture, or language—that could sustain a program of research. In this way, more general questions about the origin, depth, or purpose of human variation could be suspended while secondary information could be gathered in the expectation that the accumulation of many different forms of data would eventually confirm the existence of the thing—race—of which these were aspects.[45]

The power of the disciplinary approach is that it can proceed in a systematic and empirical way even before larger and more speculative questions receive an answer. The weakness of the disciplinary approach is that it can grind out data and information about nonexistent objects. The history of the concept of race bears witness to both.

In the end—the end of the nineteenth century—the sheer mass of statistics and other facts being accumulated overwhelmed all efforts to synthesize and organize them into a coherent understanding. While the prospect of a science of race had driven the development of disciplines, the failure of scientists to identify the essence or even the physical site of race exposed its phantasmatic nature. Confronted with this failure, scientists and others were left to face the effects of their reification of the concept and consider the ends it—and they—had served. At century's end, many of them, including several considered to be leaders in the scientific community, concluded that race, having taken the form of a scientific problem, had contributed to and arguably worsened a number of nonscientific problems, and had become a problem for science itself. A fully scientific science, having shed its former philosophical commitments, theological anchorage, and polymathic aspirations, could only be embarrassed by a continued investment in the subject of race.

But that is the end of the story, and we should begin at, or near, the beginning.

CHAPTER TWO

One or Many?

BLUMENBACH / MEINERS

The Science and Fiction of Race

At his death he would be lauded as one of the leading intellectuals in Europe, an "almost universal scholar, philosopher and sage; a naturalist, who had the glory, or rather the good fortune, of making natural history the means of proclaiming the noblest and, without doubt, the highest truth that natural history ever had proclaimed, *the Physical Unity*, and through the *physical unity* the *moral unity*, of the human race."[1] The unity of the species and the fundamental equality of all humankind were for him observed realities, scientific facts, philosophical commitments, and moral principles. On these subjects he was the most informed, progressive, and enlightened man of his time.

Nor was he merely a humanitarian. A vigorous opponent of slavery and all forms of dominance and exploitation, he grounded his convictions in both morality and science. Indeed, as the author of the memoir quoted above wrote, Johann Friedrich Blumenbach's work and example in pursuing the "*experimental method*" constituted such a compelling argument for science itself that it could be said that Blumenbach was responsible for having "made the modern human intellect what it is."[2] To such a man—humane, philanthropic, and cosmopolitan; a man described as gentle, whimsical, reticent, kindly, and ironic—the very term race was offensive. The suggestion of biological divisions within the human community was, he felt, scientifically dubious and morally repugnant. In much of his work, he would not use the word at all.

In time, however, Blumenbach would become known as the decisive figure in the emergence of a science of race that would eventually be invoked by eugenicists and race-state ideologues not excluding the Nazis. He would be implicated in theories of racial ranking and in practices of slavery and the sexual trafficking of women in faraway countries. Included in his intellectual progeny would be those who, collectively, were implicated in millions of deaths and incalculable suffering. Recently, this second reputation seems to have eclipsed the first; in 2020, a bust of Blumenbach in the institute at the University of Göttingen that bears his name was removed from its pedestal by students protesting racism.[3]

How did Blumenbach, or his reputation, take such a dramatic and complete fall; how did his reputation manage to invert itself? Have previously unpublished texts recently come to light? Have we become more insightful about the true character of his achievement? Or by a logic difficult to reproduce or even imagine, does the humane rationalist and light of the age somehow grade insensibly into the master theorist of mass slaughter so that there is no clean line between them?

To address such questions, we need to return to the time and place of the beginning.

That would be 1760 in the office of a physician in Blumenbach's hometown of Gotha, the ancestral home of the house of Saxe-Coburg. A friend of the Blumenbach family, this physician possessed a human skeleton that had so fascinated the ten-year-old Blumenbach—undoubtedly a peculiar child—that he had taken to visiting the office at times when he knew the doctor would be out so he could spend hours alone with the bones. Memorizing the ensemble, he set out to create a copy in his own home so he could devote even more time to the study of comparative osteology. "For this purpose," his memoirist writes, the precocious lad "made frequent journeys in the night to the cemeteries." Unable to meet all his needs by this means, he was forced "to content himself with the bones of our domestic animals."[4]

These he assembled in his bedroom, where he shut himself up with his treasures and, with a constancy of interest that belied his years, gave himself over to self-guided study until the day when a servant entered his room and, on beholding "that ingenious commencement of a *human skeleton* . . . cried out sacrilege and scandal." The assemblage was hastily removed to the loft. "Such," the memoirist adds, "was the modest beginning of the famous collection whose reputation has become universal." The hoard ultimately consisted of 245 skulls and fragments, along with many other body parts including the hair of albinos and two mummies—a veritable "Golgotha," as Blumenbach described it in later years.[5]

Blumenbach was, then, already a collector as well as a recognized prodigy when, at the age of twenty, he went to the University at Göttingen, where he made the acquaintance of an elderly professor. Ignored by students and forgotten by everyone else, this learned gentleman had in his possession a remarkable collection of books—geography, philology, voyages, curious objects, and pictures of life in distant nations. He shared his materials with the young man, who was often the sole audience for his wide-ranging reflections on a lifetime of heterodox learning about the peoples of the world. On the professor's death, the protégé became the curator of this hoard, and even assembled his own collections not only of human bones but of weapons, insects, books, worms, paintings, medals, mummies, drawings of odd body parts, and vegetables.[6] But the immediate result of this singular tutelage was a dissertation published in Latin in 1775 as *De Generis Humani Varietate Nativa* (*On the Natural Variety of Mankind*) when the author was twenty-three.[7]

In this remarkable work, the young scholar proposed, as a scheme "serviceable to the memory," four "classes into which the men of our planet may be divided": the European, Asian, Ethiopian, and American (NV, 1:100). He proposed these not as biological categories but as generic variations, forms taken by the human species. In the third, heavily revised edition published twenty years later, a fifth group, the Malay, would be added. This work would conclude with the clearest possible formulation of the conviction that anchored Blumenbach's life's work from beginning to end: that humankind, while various, was—as he put it in the final sentence—"*one and the same species*" (NV, 3:276).

While containing elements that we would now identify as theological, philosophical, aesthetic, and even fabulistic, *Natural Variety* became a milestone in the history of science, and a foundational text for the field that became known as bioanthropology or physical anthropology. In the eyes of some, this book, particularly the heavily revised and even reconceived third edition published in 1795, was primarily responsible for the creation of a naturalistic science of humanity, the field in which the concept of race would eventually become a central concern.[8]

The book's influence extended well beyond a single field. Thinkers of various kinds, including David Hume, George-Louis Leclerc (Comte de Buffon), Adam Smith, Lord Kames, Adam Ferguson, and Immanuel Kant had tried to create a naturalistic philosophy that would be compatible with the methods of empirical science, but with the formidable exception of Buffon, none was a scientist to compare with Blumenbach. By the time the third edition appeared, Blumenbach had become the author of definitive textbooks on physiology, natural history, and comparative anatomy. He had performed vivisections on living animals, discovered how

FIGURE 2.1. Johann Friedrich Blumenbach, shown next to the skull of Robert Bruce, king of Scotland; portrait by Ludwig Emil Grimm, 1823. Public domain.

to freeze quicksilver, dissected the eye of a seal, examined the teeth of mummies, grown a new eye in a water lizard, experimented with gases on various animals, expounded on the skeleton of "the dead female *Cercopithecus*," outlined the function of the intermaxillary bone on the two-horned African rhinoceros, given the woolly mammoth its scientific name, and penned a judicious critique of Xenocrates's *On the Aliment in Aquatic Animals*.

A scientist of impeccable credentials, Blumenbach had practiced and conceptualized his discipline of physiology so fully that in his hands it almost qualified as a philosophy and even an ethic. The cumulative result of Blumenbach's work on the human species, his memoirist claims, could be described in philosophical and even theological terms as the restoration of original unity to a human race that had been in danger of forgetting it: "The word Humanity has its whole effect in Blumenbach alone."[9]

A majestic career arc—until, as previously noted, it wasn't.

To understand the totality, we need to go back not to the beginning but to a time just before the beginning, when the Swedish botanist Carl von Linné, more commonly known as Linnaeus, was publishing his majestic taxonomy of creation, *Systema Naturae*, translated as *A General System of Nature*.[10] Writing, like Blumenbach, in Latin as the language of learning, Linnaeus established the categories that governed the "three grand kingdoms," as his subtitle put it, of the natural order, boldly placing humanity not near to God as the master of all creation but as one among many species in the "Animal Kingdom."

Every kind of organism in the natural system developed by Linnaeus occupied a unique niche, which he identified by a binomial designation indicating its place in the larger order. The species *Homo sapiens*, as Linnaeus called humans, was one of two in the genus *Homo*, the other being "monsters," of which there were six varieties. Linnaeus defined the subgroups of *Homo* by the four geographic regions in which they were to be found, adding the further dimensions of color, Hippocratic temperaments or "humours," and various cultural traits or characteristic attitudes. Both humans and monsters were grouped together with the genus of *Simia*—apes, monkeys, lemurs, and bats, all characterized by their cutting foreteeth and "lactiferous teats"—in the order of primates, in the mammalian class, in the animal kingdom.

On purely physical grounds, there was, Linnaeus implied, little to distinguish humans from apes; indeed, he told a friend that he could not discover the difference between humans and orangoutangs.[11] Nor did diurnal habits and bipedal locomotion really indicate the distinctive nature of *Homo sapiens*. It was the unique ability of the species (which included an outlier group of four-footed, mute, and hairy "wild men") to form a consciousness of its own existence that truly set humanity apart. Linnaeus translated this capacity into an imperative shared by no other group: *Nosce te ipsum*, Know Thyself.

Overcoming early outrage from some quarters at the fact that the lineage of Adam and Eve had been degraded by being incorporated into the natural world over which God had granted them dominion, the scientific community had celebrated Linnaeus's phenomenal achievement—a taxonomy ultimately embracing ten thousand organisms, each representing what Linnaeus considered a separate thought in the mind of God.

One aspect of this achievement was, however, only revealed with the passing of time. By positioning humanity in the natural world, Linnaeus invited speculation about divisions within the human like those observable in other organisms. In the case of other animals, questions about

MAMMALIA.

ORDER I. PRIMATES.

Fore-teeth cutting; upper 4, parallel; teats 2 pectoral.

1. HOMO.

Sapiens. Diurnal; varying by education and ſituation.

2. Four-footed, mute, hairy. *Wild Man.*
3. Copper-coloured, choleric, erect. *American.*
   Hair black, ſtraight, thick; *noſtrils* wide, *face* harſh; *beard* ſcanty; *obſtinate*, content free. *Paints* himſelf with fine red lines. *Regulated* by cuſtoms.
4. Fair, ſanguine, brawny. *European.*
   Hair yellow, brown, flowing; *eyes* blue; *gentle*, acute, inventive. *Covered* with cloſe veſtments. *Governed* by laws.
5. Sooty, melancholy, rigid. *Aſiatic.*
   Hair black; *eyes* dark; *ſevere*, haughty, covetous. *Covered* with looſe garments. *Governed* by opinions.
6. Black, phlegmatic, relaxed. *African.*
   Hair black, frizzled; *ſkin* ſilky; *noſe* flat; *lips* tumid; *crafty*, indolent, negligent. *Anoints* himſelf with greaſe. *Governed* by caprice.

Monſtroſus Varying by climate or art.

1. Small, active, timid. *Mountaineer.*
2. Large, indolent. *Patagonian.*
3. Leſs fertile. *Hottentot.*
4. Beardleſs. *American.*
5. Head conic. *Chineſe.*
6. Head flattened. *Canadian.*

The anatomical, phyſiological, natural, moral, civil and ſocial hiſtories of man, are beſt deſcribed by their reſpective writers.

Vol. I.—C 2. SIMIA.

FIGURE 2.2. From Sir Charles Linné (Linnaeus), *A General System of Nature*, trans. William Turton (1806). Photo: Alamy.

the nature, depth, and purposes of such divisions had a purely taxonomic interest. In the case of human beings, these questions reached deep into human self-understanding, leaving little about the human condition untouched. In the years that followed, attempts to "Know Thyself" would often take the form of puzzling out such questions under the heading of a concept unavailable to Linnaeus, the concept of race.

In the early work of Blumenbach, we see him edging toward such an inquiry. As a young man, he largely shared Linnaeus's general view, but felt there was something deeply imperfect about the whole thing.[12] In the first edition of *Natural Variety*, he accepted, with modifications, the four geography-based categories, but quietly eliminated any discussion of temperaments or humors, and focused entirely on biologically significant features. Linnaeus had, for example, described Americans as choleric, erect, and "regulated by customs," Asians as "melancholic," "rigid," and "governed by opinion," and Africans as "phlegmatic," "relaxed," and "governed by caprice" (or in an early edition, "by the arbitrary will of the master"); Europeans were "fair," "brawny," and "governed by laws." Such considerations may have seemed to Linnaeus a legitimate part of science, but Blumenbach found them irrelevant and eliminated them along with the wild men and the monsters.

While endorsing the idea that humanity belonged in the natural world, Blumenbach insisted on a sharper distinction between primates and humans, who were, he pointed out in the 1795 edition, "lord and master of the rest of the animals," a fact indicated by a host of distinctive features in *Homo sapiens* that Linnaeus had not noticed or credited: few and feeble instincts, a sense of the future, an upright walking position, the use of two hands, the position of the heart, the extreme delicacy of the mucous membrane, omniverousness, extraordinary adaptability, the lack of bodily hair, nocturnal "pollutions" in men and "menstruous flux" in women, various features of the brain and vertebrae, long infancy, late puberty, susceptibility to certain diseases such as hysteria and nymphomania, toothaches, intestinal worms, the ability to copulate whenever appetite and opportunity coincided, an extended intermaxillary bone, the ability to laugh and cry, belching—and above all, of course, "*the use of reason*" (NV, 3:180–87). All these constituted an integrated set of physical and moral barriers standing between humans and primates. This distinction, Blumenbach said, was a crucial element in the "highest work of the Supreme Being," who had, with sublime clarity, created clean categories and clear distinctions among species (NV, 1:133).

Blumenbach laid particular emphasis on a point that had for him a more than theoretical importance, the equal potentiality of all humans, differences notwithstanding. He based his conviction on research, or such research as he could conduct from his home and the university library, with travel narratives standing in for the worldly experience he would never acquire. He had also taken the opportunity provided by the markets of Göttingen to study people from other lands. Everything he saw and read strengthened his conviction of a fundamental human unity. Describing in one book a series of "exemplary lives," he detailed the remarkable achievements of men who had grown up in cultures with no writing systems, no educational institutions, and no tradition of urbanized living, as a way of dispelling any beliefs people might have about the natural limitations of some human varieties.[13]

Blumenbach was particularly interested in Negroes, as people of African origin were called in the English translation of his work.[14] He accumulated a library of Negro authors including Phyllis Wheatley, and corresponded with and even met some including Olaudah Equiano. And as early as 1790, in the first edition of a treatise called *Contributions to Natural History*, he included an extraordinary eight-page section titled "Of the Negro in Particular" in which he insisted, against a belief commonly held, that these people, too, were made in God's image. He gave sympathetic and appreciative descriptions of people he knew, condemned the slave trade, referred to Negroes as "*our black brothers*," and argued strongly against any suggestion that Negroes might be different in intellectual capacity or moral sensibility from any other group. He concluded that "there is no so-called savage nation known under the sun which has so much distinguished itself by such examples of perfectibility and original capacity for scientific culture, and thereby attached itself so closely to the most civilized nations of the earth, *as the Negro*."[15] There was almost nothing like this in contemporary scientific literature and would not be for many years.[16]

Blumenbach's empirical approach to the question of the human difference in the order of creation began with the genitals, of which he was a lifelong student. As a young man, he made detailed observations of the genitals of a female elephant that had been exhibited in Germany. And when, on a trip to Holland in 1791, he heard that a whale, "a specimen of the *Balæna boops* [fin whale] about fifty-two feet in length," had been thrown up on the shore, he rushed to the site so he could inspect the monster's prodigious clitoris (NV, 3:171). Female genitals in particular had for Blumenbach a moral as well as taxonomic significance; or to be more precise, the taxonomic difference reflected a moral difference. In the first edition of *Natural Variety*, he noted that the hymen, as "the guardian of chastity, is adapted to man who is alone endowed with reason; but the clitoris, the obscene organ of brute pleasure, is given to beasts also" (NV, 1:90).

Insensitive and indifferent to the taxonomic significance of hymens, Linnaeus had, Blumenbach believed, pointed in the wrong direction. In fact, he had pointed in two wrong directions, toward an indistinct boundary between primates and humans on the one hand and toward quite distinct boundaries separating the subspecies of the human on the other. While never doubting that *Homo sapiens* was one species, Linnaeus had nevertheless presented his various subspecies as products of geography, with groups as distinct from each other as the continents. The young Blumenbach found this premise offensive to the principle of the unity of the human species. He repeatedly reminded his reader that the features of the various groups "all do so run into one another [and] one variety of mankind does so sensibly pass into the other, that you cannot mark out the limits between them" (NV, 1:98–99). "There is," he wrote, "an almost Ωman lady through the yellow, the red, and the dark nations, to the Ethiopian of the very deepest black" (NV, 1:107).

If to Blumenbach's mind the boundary lines among human varieties were indistinct and the categories themselves arbitrary inventions, the divisions between species in the natural world were bright, clear, and enforced by nature. According to a principle established by Buffon and endorsed by Kant, the test of species identity was the ability to procreate: different varieties within a species could be freely fertile, while different species met with resistance. The fact that horses might mate with donkeys to produce mules while mules could not reproduce themselves suggested a rejection by nature of the mule itself, and proof that horses and donkeys were distinct categories. Of course, some animals were indifferent to nature's preferences. On occasion, Blumenbach conceded, "male brutes when burning with desire, and unable to obtain females of their own species," may perchance copulate with females outside their species, but these incidents do not generally result in offspring, and for good reason: God, working through nature, had not contemplated such creatures and did not support their continuance (NV, 1:75).

Divine disapproval of unsanctioned minglings was often indicated by genitalia that while "providently and carefully adapted for copulation" within a species, were so positioned or sized in "distant genera" as to "render the whole thing impossible" (NV, 1:75–76). To be sure, the perverse ingenuity of animals in pursuit of *mésalliances* could never be underestimated. A "*Gallus calecuticus* [rooster] has been known to tread the duck," a "credulous antiquity" believed in human-brute hybrids, and men have on occasion "very wickedly had connexion with beasts" (NV, 1:76). But in all such cases, "the Supreme Being foresaw these disgusting kind of unions and took care to render them futile" (NV, 1:80). Nor could travelers' tales of "lascivious male apes" attacking women and producing offspring be credited, "since those very travellers relate that the women perish miserably in the brutal embraces of their ravishers" (NV, 1:81).

Registering a species revulsion at the spectacle of creatures endowed with both hymens and reason trafficking in the clitoral world of animality, Blumenbach announces that he has reached the limit of his tolerance: "I now leave this disgusting theme" (NV, 1:81).[17]

Blumenbach's eagerness to be rid of the theme of cross-species fornication reflects not only his scientific understanding of the threat to nature presented by illegitimate mixtures but also his professional sense that a scientific dissertation was not the place for a recitation of barnyard irregularities or the misguided beliefs of a prescientific era. Like nature itself, science—a term not yet fully hardened; the word scientist did not appear until 1834—depended on clarity, consistency, and the preservation of distinctions. One of the most important such distinctions concerned rational discourse as such. If people once believed in human-animal hybrids, it was because they had been misled by "fables," and one of the tasks of the modern intellect is therefore "to separate the stories from the truth," discursive species as distinct from each other as roosters from ducks (NV, 1:132).

And yet there was a genuine problem: How to describe nature's invisible processes, immaterial entities, and abstract concepts in a way that would be comprehensible to the common reader? In order to communicate discoveries to a broader public, the scientist was forced to deploy nonscientific approximations that would accommodate the unfamiliar to the familiar, but such accommodations inevitably introduced the potential for falsehood. Struggling against his rationalist distrust of any "metaphorical and allegorical amusements" in scientific description—a distrust felt by countless scientists and not only scientists down to the present—Blumenbach reluctantly conceded the necessity of speaking in the vernacular to a public still unfamiliar with and resistant to the modes, methods, concepts, and vocabulary of the modern era. Indeed, scientists themselves struggled to articulate their concepts. "I almost despair," he said in the

1795 edition, "of being able to deduce any notion of species in the study of zoology, except from *analogy* and resemblance" (NV, 3:190).[18]

The invention, in describing categories of humanity, of a blended discourse in which the literal and the metaphorical, the serious and the merely amusing, the objective and the aesthetic were so interfused that you could not mark out the limits between them was to be Blumenbach's most enduring, if entirely accidental and indeed unintended, unvalued, and undesired contribution to the discourse on race.

When he wrote of his despair in 1795, Blumenbach may have been thinking of the manner in which he had begun the first edition of *Natural Variety* twenty years earlier, when he had launched his argument and his career with an analogy. "As I am going to write about the natural variety of mankind," he had written in the first sentence, "I think it worth while to begin from the beginning, that is, with the process of generation itself (NV, 1:69). Like many since the time of Aristotle, Blumenbach chose chickens as his example because the daily production of eggs could be taken as a compressed version of an operation that would, in other organisms, be more difficult to study: generation as such, generation as a purely biological process.

This process, Blumenbach says, is entirely centered in the hen, who "supplies the true stamina of the future fœtus." In the beginning, unity reigns, with chicken, egg, yolk, and fetus forming "but one continuous body"; at the end, an egg issues forth, each like the others. Taking chickens as the model for generation, Blumenbach implies that a species proceeds from like to like. On this basis he rests his claim that human differences—which would eventually take the form of groupings that would eventually be called races—are mere external variations, for the process of generation, left to its own devices, would produce a posterity that would in principle continue "for ever like their first parents" (NV, 1:70).

The young Blumenbach may have felt that this ingenious gambit, making human unity a downstream effect of generation, provided brilliant support for his larger thesis. The analogy is, however, imperfect, for the daily production of identical eggs results in a procession of biological dead ends, and only becomes a model for generation with the advent of the rooster, whose seed disturbs the dreamy intestinal confusion of chicken and egg, transforming sterile identity into productive difference. Wishing to insist on identity, Blumenbach describes the agency of the rooster in the most noninvasive terms possible as the alteration of the egg "in some of its parts by the influence of the male" (NV, 1:70). But the part that was altered was the crucial part, and the alteration was total, a transformation from certain death to life. Intended as a biological analogy for the zoological fact of human unity, the production of eggs actually suggests a different and deeper argument than the one Blumenbach intended, the necessity of difference in the process of generation.

One can see why Blumenbach was distrustful of analogies, which invariably introduced distraction and sacrificed scientific exactitude. But he appears to have been unable to describe the process of generation without them. Take, for example, his account of the attribute that makes generation possible, the "primeval germ." More property than entity, the primeval germ is, in Blumenbach's account, an organic inclination, a principle of passive responsiveness and potentiality within the unfertilized yolk that would, he says, "lie buried as it were in eternal slumber, were it not aroused by the access and stimulus of the fertilizing seed of the male, and particularly by the subtle odour of his parts" (NV, 1:70). What, exactly, is the primeval germ—what is he talking about? The analogy seems to have eclipsed the thing itself, for it has been rendered in luridly human terms and endowed with adult attributes and appetites—sleeping without a brain, smelling without a nose, yearning without a body.

This steamy account of the primeval germ may have caused the mature Blumenbach some discomfort not only because it humanized a mere germ but also because it so clearly evoked a "primeval" folktale. The image of the inertial woman awaiting her prince was itself the germ of the chivalric romance that became the story of Sleeping Beauty. This image made its first appearance around 1340 in *Perceforest*, an anonymous, million-word Old French prose romance, and reappeared in various forms before being taken up as a Mother Goose story in Charles Perrault's *Histoires ou contes du temp passé, avec des moralités* (1695). Blumenbach wanted to write of scientific discoveries in the present, not recast old wives' tales, especially those in which the bestial clitoris might be seen to prevail over the human hymen. Whatever the reason, the primeval germ—the essence of generation in 1775—did not survive; there is no mention of it in the 1795 text.

But other nonscientific amusements survived and even flourished in the 1795 version, among them apocrypha, ancient tales, and folk wisdom drawn from Blumenbach's lifelong reading of long-forgotten and obsolete works. Perhaps the most telling example of his dependence on narrative and even fabulistic models for biochemical processes is his treatment of the means by which the differences among groups arise. This process, which Blumenbach called degeneration, is the primary focus of his mature rethinking and therefore qualifies as the baseline concept of the theory of race that developed over the following century. The account of degeneration should represent a triumph of the "experimental method" for which Blumenbach became famous, the purest instance of "the modern human intellect" he supposedly created. Does it?

The theory can be explained in simple terms. According to Blumenbach, the causes of variation are all external to the organism. We vary from each other not because we were made that way but because we became that way. Differences in skin color, hair, and stature that some might take as signs of fundamental or intrinsic differences all have their origin in diet, mode of life, hybridism, and particularly climate.[19] Over time, the outside works its way in, producing variations that eventually reproduce themselves as it were naturally.

The principle is simple, but the various explanations are not, and not at all what would today be called scientific. Take for example the attempt in the 1790 *Contributions to Natural History* in which Blumenbach, noting that domesticated pigs exhibit great diversity while wild boars living in a single environment all look the same, argues that human diversity can be attributed to the fact that humanity was never in a wild state. Apparently unwilling to leave this striking claim altogether unsupported, he added an astonishing footnote in which he cited without comment a "very profound psychologist" who had declared that "there must have been in the primitive world a class of higher existences on earth, to whom man acted as a sort of domestic animal."[20]

A more scientifically serious attempt to explain diversity without recourse to analogy is mounted in the 1795 *Natural Variety*, where Blumenbach attempts to account for the variation of skin color among the peoples of the world, using as his example the "pitchy blackness" of the Ethiopians (NV, 3:209). Framed in medicalese as a laboratory discovery achieved by means unavailable to the public, the account is by twenty-first-century standards completely fantastical. In fact, it is more deeply fantastical than the theory of the humors proposed by Hippocrates that it seeks to supplant because it retells that (uncited) ancient tale in the vocabulary of science, as though modernity had done no more than discover and redescribe the mechanisms intuited by the ancients.[21] (*Warning to reader: gird your loins, even for this greatly simplified, reader-friendly redaction.*)

The tale begins in Ethiopia, or rather with the air above the land of Ethiopia, which has distinctive "gasiform constituents, the accessories of light, heat, electricity, &c," in their varying proportions and intensities as determined by topography, elevation, vegetation, proximity to water, and so forth (NV, 3:196). Through a complex process, climate is converted to chemistry. The carbon-rich air engulfs and bathes the body on the outside, penetrating to the middle layer of the epidermis, the Malpighian mucus, which becomes in the Ethiopian singularly spongy, thick, and black, whereas in other peoples it is white. But since the chemical elements in the air are sucked into the living laboratory of the body from birth, they enter the skin from the inside as well, where they are distributed by the arterial blood. Some elements wind up once again on the peripheral integuments of the body where they come into contact with the aforementioned Ethiopian air, while other elements are carried back by the venous blood to the respiratory organs, where they produce modifications in the "humours, especially of fat, bile &c" (NV, 3:197). The "sanguineous vessels of the corium" play, of course, a key role in storing and secreting the carbon, which is "in greater quantity in the atrabilious." The connection between the "manufactory of the bile" and the "common integuments" hardly needs to be emphasized. Thus we can see how pitchy blackness is "not congenital even to the Ethiopians themselves, but is acquired by the access of the external air" (NV, 3:211). In short, Ethiopians inhale their color.

"How the Ethiopians Got Their Blackness"—a delightful fable, to be placed in the family library alongside the other great just-so story of Ethiopia, Rudyard Kipling's "How the Leopard Got His Spots."

Blumenbach's most famous contribution to the lexicon of science, the *Bildungstrieb*, is rendered in even more directly analogical terms. Variously translated as "formative force," "formative impulse," or "formative energy," the *Bildungstrieb* is conceived as a powerfully effective but mysterious agency operating within the organism to develop embryos, heal wounds, and create offspring similar to the parents. Given all those responsibilities, one might think that the formative force would be everywhere, and everywhere observable. It is not. It is in fact invisible, and knowable only through its effects. "The term *Formative Impulse*," Blumenbach says, "of itself, explains nothing," and serves only "to designate a peculiar power . . . a power, the constant agency of which we ascertain by experience, whilst its cause . . . still remains, in the strictest sense of the word—'*qualitas occulta*'" [hidden quality].[22] In the 1795 *Natural Variety*, he repeats that the formative force was not so much the cause of variation "as some kind of perpetual and invariably consistent effect, deduced *à posteriori*, as they say. . . . Just in the same way as we use the name of attraction or gravity to denote certain forces, the *causes* of which however still remain hid, as they say, in Cimmerian darkness" (NV, 3:194).

"Cimmerian darkness," a reference to a nomadic people living in a land of perpetual mist in Homer's *Odyssey*, is a literary analogy of the kind Blumenbach found necessary for expository purposes; but so is concealed agency, which could be considered a generic feature of the story Blumenbach wanted to tell. Inferred from its effects rather than directly observed, the *Bildungstrieb* is a *nom de science* for a type of agency perfectly suited to literary treatment: a figure of masked intrigue, a secret agent, a biochemical Scarlet Pimpernel operating in the shadows to create the familiar world we inhabit and take for granted.

It would be wrong to think that the erudite Blumenbach unconsciously interpolated specific images, conventions, or tales. What we actually see in his labored accounts of biochemical processes is an insistent blurring of the line we might imagine to exist between scientific description and narrative. As we have seen, he regretted but tolerated the imprecision introduced by analogies as the cost of intelligibility; but surely the deeper determinant was his commitment to Enlightenment humanism. Blumenbach objected to rival accounts of generation in which all organisms were, at the dawn of creation, mere "undeveloped germs" that had developed in various directions, because these accounts implied a world of meaningless process. His objection to the "vulgar system of evolution" being bruited about in the 1790s was not only that it was unscientific but that the world it described was insulting to God in that it accorded no dignity to the divinely ordained system of nature or to the humans who stood at its apex (NV, 3:194). He wanted an account of human variation that was less mechanical, more purposive, more aligned with the Enlightenment sense of a progress toward perfection. He intuitively understood that the empirical method was inadequate for this purpose, and that only analogical narrative could provide the sense of wholeness, meaning, direction, and value appropriate to the grand spectacle of generation.

The problem Blumenbach confronted was how to think about and represent things that both do and do not exist. The problem begins with entities such as the primeval germ and the formative force that can only be represented by analogy with persons, but these are merely local instances of a larger phenomenon. Categories of human variations, to which, as we will see, Blumenbach gives names and attributes, display the same ambivalence at a higher level: obvious to the eye, they have, as he insists, no basis in physiological reality—even the concept of the category is a metaphor. Within the science of the primeval germ resided a primeval germ of analogy.

Many scientific works then as now include what we would consider nonscientific elements, but in Blumenbach's case, his subject—the real-unreal varieties of mankind—can only be represented in a mongrel or mulatto discourse in which attributes of imaginative literature and empiricism are so thoroughly ambiguated that what now seem to be opposed ways of thinking run into one another so that you cannot mark out the limits between them.

Blumenbach is widely acknowledged to be a pioneer in the study of race even though his terms and distinctions were not retained and nobody in subsequent years continued his particular kind of work. He is actually a pioneer in the engagement with the problem of describing categories of humanity for which there is no basis in nature. In a sense, he is a tragic figure. If he had stuck with his initial conviction that humanity was a single species with no natural divisions, he might today be hailed as an antiracist Enlightenment hero. If he had merely outlined the process of degeneration, speculated about the origin of variations, and posited something called a formative force, he might be remembered as a brilliant but quixotic theorist, a visionary of sorts; his bust might still be on its pedestal at Göttingen. But following an imperative he seemed powerless to resist, he devoted himself instead to a description of the categories of human variation, a contribution he intended to be "serviceable to the memory" that has indeed been remembered—to the great detriment of his reputation.

Looking Beneath the Racial Face

Recognizing the limitations of Linnaeus's mighty effort, Blumenbach still understood the utility and appeal of a physical taxonomy of human groups. But how could categories be identified in a unified species that exhibited a regular gradation of forms? Neither skin color nor hair nor skeletal structure provided any clear marker of difference since one variation slid insensibly into the next. At first Blumenbach was intrigued by the "racial face," which, he thought, might exhibit enough constancy of form to support generalizations. This would be a vast and agreeable field for research, with Swiss, Savoyards, Turks, and others having well-known characteristic facial expressions that might be describable in scientific terms. The "goat-like" face of the Jews, for example, had been noted by artist Benjamin West, who linked it to "the transit and conflux of the septum" between the nostrils and upper lip (NV, 3:170). But Blumenbach quickly decided against this approach, noting that some people who are sanguine and guileless in youth grow morose and melancholy with age, and that the faces of people who age, suffer, or move to different locations change with the circumstances. Relocated New Hollanders, for example, no longer beset by the gnats that afflicted them in their homeland, lose the irritable expression that might have been thought to define them (NV, 3:232). The frequently reported tendency of the people of a given region to come to resemble the local animals was another complicating if dubious factor.

No, Blumenbach decided; the face was too malleable, too direct a function of climate and contingency, to be a source of scientific interest. For harder evidence, the scientist must look elsewhere. He began to suspect that the treasure had been in his possession all along, concealed beneath the racial face. "*Above all*," he said in his third edition, the scientist should look to "*the shape of the skull*" (NV, 3:193).

As we have seen, Blumenbach's interest in skulls extended back to his boyhood, when, however, he had been unable to develop a sophisticated research project. As his scientific knowledge increased and his collection grew, he began to suspect that the eighty-two skulls in his possession might exhibit some regularities worth studying. Then, in 1791, a publication by the Dutch polymath Petrus Camper appeared that seemed to offer a way forward for scientific inquiry.

The text was a lecture that had been presented to the Amsterdam Drawing Academy in 1770 as a way of helping artists make better portraits. Published posthumously by Camper's son under the title of *Dissertation sur les variétés naturelles qui caractérisent la physionomie des hommes des divers climats et des différens âges* (Dissertation on the natural varieties that characterize the physiognomy of men of diverse climates and of different ages), the lecture outlined a new way of measuring faces: one could represent the cranial form on paper, then draw two lines, one horizontally from the base of the nostril to the ear and the other perpendicularly from the upper jawbone to the most prominent part of the forehead. The result would be the "facial angle."

This small drawing technique would prove to be an astoundingly productive if altogether misleading innovation—not in artistic pedagogy but in the study of human variation.

Although his lecture did not concern the origins of variation, Camper had definite views on the subject that were much like those of Blumenbach. Both were monogenists who believed in the single creation of a unified human species, with variations attributable to external causes. Like Blumenbach, Camper was contemptuous of those who believed that Negroes were the issue of whites and primates, and rejected any suggestion of a hierarchy of races. But Camper argued these cases elsewhere; in this text, he merely considered the shapes of heads, and his visualizations lent themselves to polygenist theories of plural creations that, had he been asked about them, he would have dismissed out of hand as being scientifically unfounded and morally offensive.[23]

In presenting the facial angle, Camper ranged his profiles from the lowest to the highest, beginning with the monkey (42°) and the orangutan (58°), followed by the "prognathous" Angolan and the Asian or "Kalmouk" (both 70°), and ending with the European's near-vertical 80°. At the far right was Camper's sketch of a sculpture from Greco-Roman antiquity, representing a perfection rarely attained in nature, a majestic 100°. While Camper's intention was to prove the naturalness of certain non-European features that many believed to be artificial results of manipulation by midwives or mothers after birth, the illustrations actually undermined the idea of a unified and distinct human species by visually implying a continuum, a temporal and developmental succession in which Africans were adjacent, and perhaps biologically akin, to primates.[24]

This was not the inference Blumenbach drew. What he saw in Camper was a limited but extraordinarily suggestive innovation in the study of human variation. He immediately recognized the possibility of studying skulls in ways that might demonstrate the workings of the formative force in producing specific variations. After completing a systematic study of his collection of specimens, he felt newly confident about the possibility of a sophisticated approach to variation through the study of skulls, publishing his preliminary and purely descriptive results in fascicles under the title of *Decas Craniorum* beginning in 1790. By 1795, he could declare with confidence that it was in the shape of the skull that we could observe "*racial form*," "*racial characters*," "*racial varieties*," and in general a "constancy of characteristics which cannot be denied" (NV, 3:234, 236, 237, 235). The object and the method required to establish a science of human variation was at last identified by the preeminent scientist of the era: through the close observation of skulls, natural science could become transformed from a mere description of appearances to a genuine taxonomy that would be far superior to that of Linnaeus.

The crucial, even fateful question was, a taxonomy of what? In one sense, Blumenbach was explicit about the fact that he was not studying natural categories, since as far as nature was concerned, humanity was one and indivisible. He was, rather, attempting a taxonomy of the visible distinctions, the "principal varieties," created by degeneration. Since "climate is the principal cause of the racial face," a study of skulls could only reveal the long-term effects of the environment in shaping the human organism (NV, 3:229). And yet he himself had confused the matter by reluctantly but repeatedly "admit[ting] the position of Hippocrates" that acquired characteristics can be inherited and take the form of "a second nature"

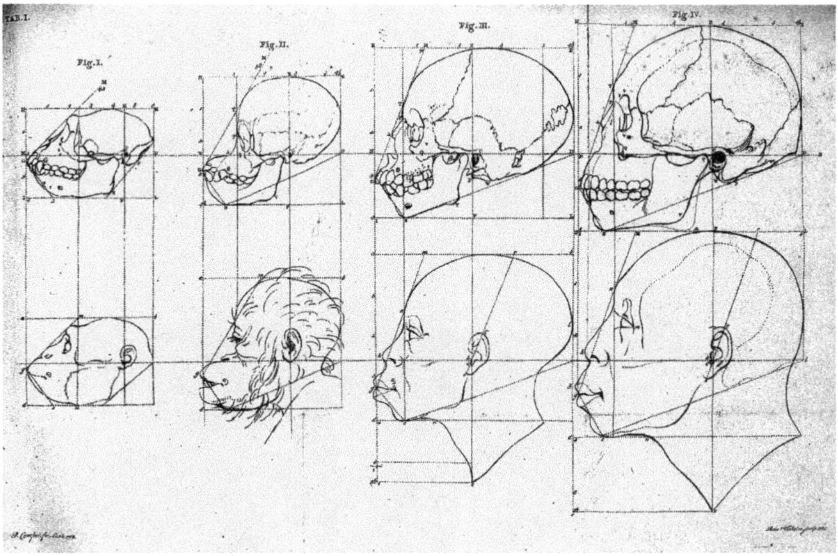

FIGURE 2.3. Petrus Camper, *Dissertation sur les variétés naturelles*, trans. H. J. Jansen (1791), 170–71. Public domain.

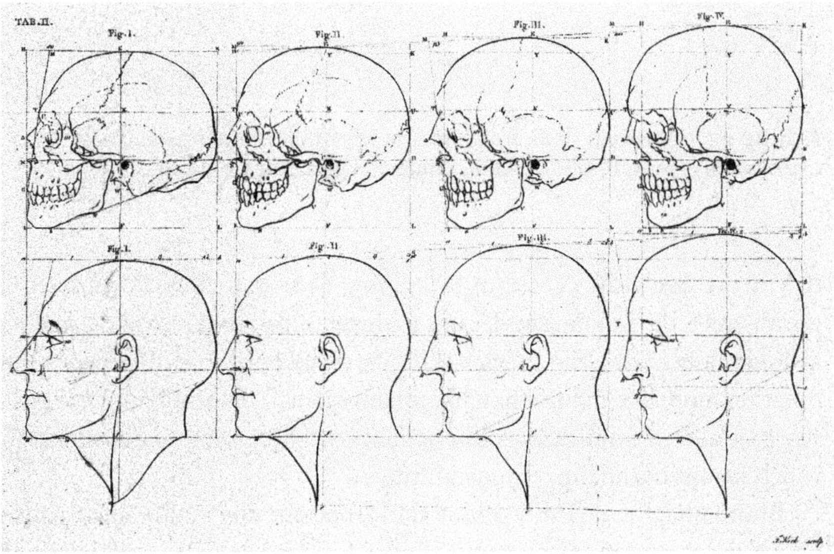

FIGURE 2.4. Petrus Camper, *Dissertation sur les variétés naturelles*, trans. H. J. Jansen (1791), 172–73. Public domain.

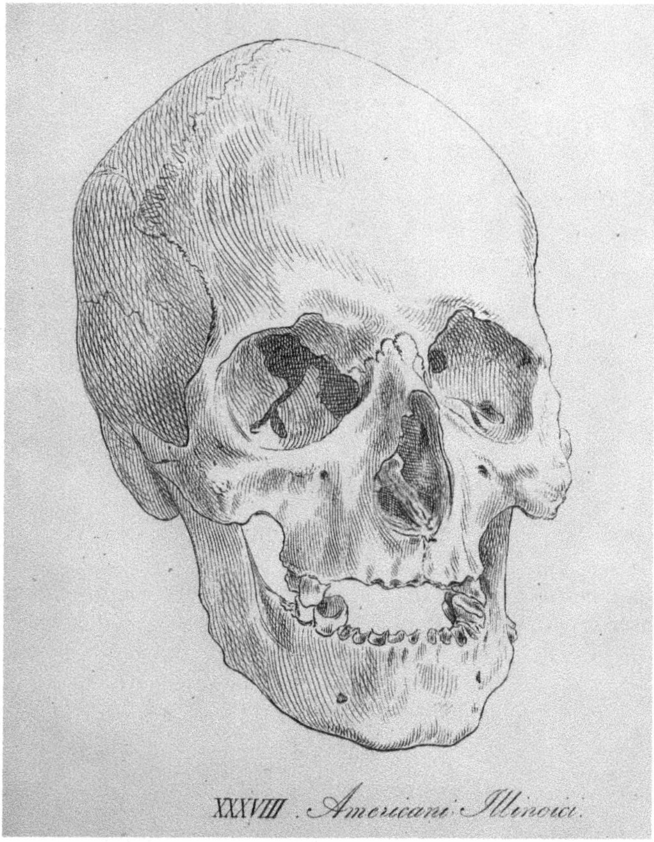

FIGURE 2.5. American skull, from Johann Friedrich Blumenbach, *Decas Craniorum* (1790). Public domain. Source: Wikimedia Commons.

(NV, 1:121. See also NV, 1:130; NV, 3:203, 242, 259, 262). Considered in the abstract, the human species has a single unified nature, but considered as it actually exists, the species exhibits many heritable differences that function more like nature than like acquisitions. This ambiguity has given rise to a host of implications Blumenbach would have repudiated but for which he has borne the responsibility.

Blumenbach's curiosity about skulls became the single most consequential innovation in the study of race, for it launched a multi-national search for the secret of racial difference in the shape and volume of crania that would last for over a century.[25] The project would, as we will see, take what now seems a deeply sinister turn, particularly in the United States; and many, looking for the origin of that turn, have accused Blumenbach of having taken it himself, setting the precedent and the course for others.

But before we consider that argument, we should ask why Blumenbach might have been so fascinated by skulls that he placed his entire theory and most cherished principles at risk by treating them as a source of evidence about human variation.

Blumenbach may have been an early adopter of the theory that skulls held the key to human variation, but he was far from the first to sense that they held a mysterious and revelatory significance for human existence. In their changeless state, skulls have long seemed to the imaginative observer to represent a comment on the pretensions, vanities, and illusions of those who, for a brief time, strut and fret their hour upon the stage of life. One's gorge may rise at the sight of them, as Hamlet's does, but they invite rumination. Indeed, one of Blumenbach's biographers says, "The curious and inquisitive of both sexes" were in the habit of visiting Blumenbach's Golgotha "to wonder and reflect."[26]

A deeply intelligent and imaginative man, Blumenbach may have been stimulated by his collection of skulls into reflections of a kind that were not only scientific but philosophical or literary, or a blend of all three. He may have been impressed by the ability of skulls to serve as both a protective wall shielding the innermost part of personhood—"the hardest possible receptacle of . . . the softest and noblest entrail"—and a source of timeless truth that rendered nearly irrelevant the brain and all its transitory thoughts (NV, 1:114). He may have known of the speech in Shakespeare's *Richard II* about how the "hollow crown that rounds the mortal temples of a king" produces an illusion of majesty and power that in fact expires with the king, while the crown itself, like the skull, survives as an enduring reminder of the brevity of life and the vanity of human pretensions. If he sensed any of these possibilities, Blumenbach may have understood the stakes of skull-study to be higher than the empirical method alone could reach.

If Blumenbach had articulated such thoughts, his reflections might have been considered to be as irrelevant to actual science as Richard's to actual kingship. He might have been seen as a deep-feeling poetical thinker and a man of fundamental goodwill. But instead of treating skulls as provocations for profundity, Blumenbach treated them as products of externally induced degeneration whose effects had become over time effectively natural in that they were reliably inherited. In Blumenbach's Latin, *degeneratio* carried no suggestion of deterioration or degradation, and meant merely difference over time; but not all his readers read Latin. He was most widely read in German or English translations, and in both these languages, degeneration had a distinctly negative implication. To readers in these languages, it could easily appear that Blumenbach was suggesting that variations represented fallings-off from an origin, deviations from a norm, or degraded versions of an ideal of primitive perfection.

Even so—even after suggesting that differences represented effectively natural categories, and after using a term that could be easily misunderstood—Blumenbach might not have been accused of being the founder of an immense intellectual movement that propagated and vindicated practices of disenfranchisement, expropriation, exploitation, and violence if he had not gone on to nominate a candidate for the undegenerated origin, or if he had not followed Camper's example in presenting a series of skulls.

But he did, he did. In the 1795 edition of his text he included two illustrations, rows of skulls from his collection that represented each of the classes into which he had, for purposes of organizational clarity, divided humanity. In the first, an "Ethiopian" is on one side, a "Reindeer Tungus" or Siberian is on the other, and a "Georgian" female is in the center. In the second, a fold-out drawing, five varieties are given generic names and placed in a line: the Asiatic, Malay, Georgian, American (Indian), and Ethiopian.

He identified his Georgian skull—the only one in his possession from that region—as a representative of the "Caucasian variety" based on the fact that Georgia is near Mount Caucasus. Not a Biblical literalist but far from being disrespectful of revealed religion, Blumenbach seemed to be alluding to the terminus of Noah's journey that the Bible refers to as "the mountains of Ararat" just south of the Lesser Caucasus Mountains. As natives of the geographical starting-place of post-Deluge humankind, "Caucasians" would, Blumenbach said, seem to be the "primeval" group, the "autochthones of mankind" (NV, 3:269).

Blumenbach adduced as supporting evidence for this hypothesis the fact that the people of Georgia are "white in color, which we may fairly assume to have been the primitive colour of mankind," since it is easier for white to become dark than the reverse, once "the secretion and precipitation of this carbonaceous pigment has once deeply struck root" (NV, 3:269). This proximity to the geographical origin of the species helped explain the striking appearance, the "beautifully symmetrical, somewhat globular" form, of the skull of the Georgian female, which had, Blumenbach reported, "that kind of appearance which, according to our opinion of symmetry, we consider most handsome and becoming" (NV, 3:237, 265). Even today, he noted, it is commonly understood that the people of

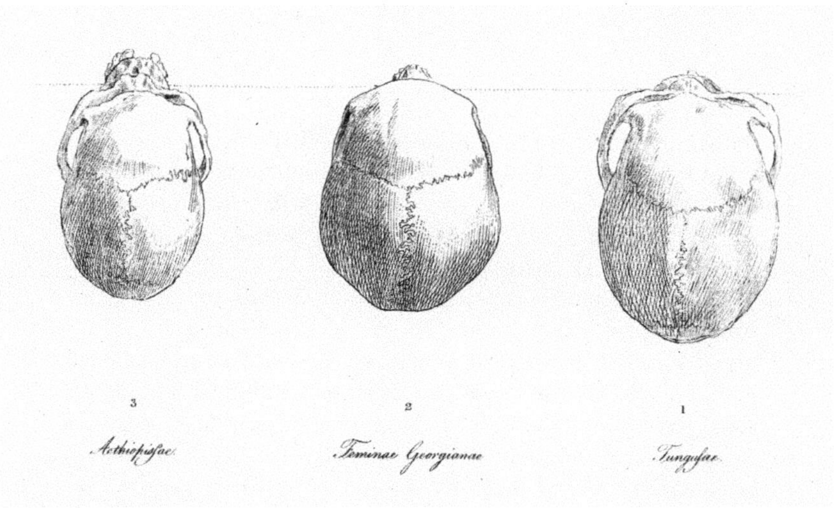

FIGURE 2.6. Johann Friedrich Blumenbach, *On the Natural Variety of Mankind* (1795), fig. 1, plate 3.

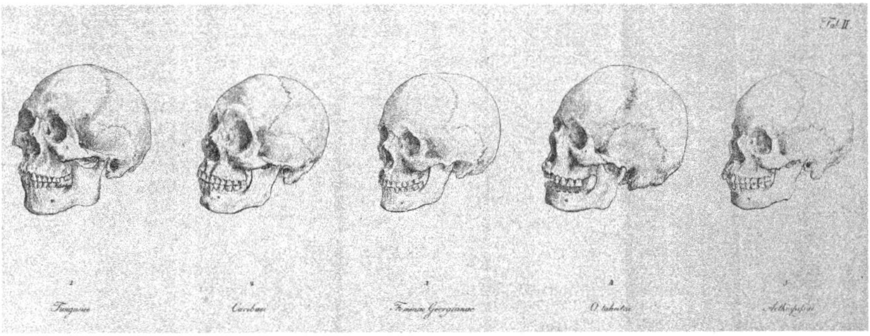

FIGURE 2.7. Johann Friedrich Blumenbach, *On the Natural Variety of Mankind* (1795), fig. 2, plate 4.

Georgia represent "the most beautiful race of men," an unsurprising fact given that, as the autochthones of mankind they are the least-degenerated of all human groups (NV, 3:269).

The page in which these claims are gathered together contains the single most consequential passage in the history of racial theorizing.

Gazing into the hollow sockets of the Georgian skull, Blumenbach may have felt he was looking directly at the origin of humankind, even holding it literally in his hands. But he then made a further, even more consequential discovery that was as triumphant for his theory as it has been damaging to his subsequent reputation, that the skull of the white autochthon, the beautiful Georgian female, resembled the European skulls in his possession. At

CAUCASIAN VARIETY. 269

return to our pentad of the varieties of mankind. I have indicated separately all and each of the characters which I attribute to them in the sections above. Now, I will string together, at the end of my little work, as a finish, some scattered notes which belong to each of them in general.

85. *Caucasian variety.* I have taken the name of this variety from Mount Caucasus, both because its neighbourhood, and especially its southern slope, produces the most beautiful race of men, I mean the Georgian[1]; and because all physiological reasons converge to this, that in that region, if anywhere, it seems we ought with the greatest probability to place the autochthones of mankind. For in the first place, that stock displays, as we have seen (s. 62), the most beautiful form of the skull, from which, as from a mean and primeval type, the others diverge by most easy gradations on both sides to the two ultimate extremes (that is, on the one side the Mongolian, on the other the Ethiopian). Besides, it is white in colour, which we may fairly assume to have been the primitive colour of mankind, since, as we have shown above (s. 45), it is very easy for that to degenerate into brown, but very much more difficult for dark to become white, when the secretion and precipitation of this carbonaceous pigment (s. 44) has once deeply struck root.

86. *Mongolian variety.* This is the same as what was formerly called, though in a vague and ambiguous way, the Tartar variety[2]; which denomination has given rise to wonderful mistakes in the study of the varieties of mankind which we are now busy about. So that Buffon and his followers, seduced by that title, have erroneously transferred to the genuine Tartars, who

---

[1] From a cloud of eye witnesses it is enough to quote one classical one, Jo. Chardin, T. I. p. m. 171. "The blood of Georgia is the best of the East, and perhaps in the world. I have not observed a single ugly face in that country, in either sex; but I have seen angelical ones. Nature has there lavished upon the women beauties which are not to be seen elsewhere. I consider it to be impossible to look at them without loving them. It would be impossible to paint more charming visages, or better figures, than those of the Georgians."

[2] On the origin of this erroneous confusion, by which the name of Tartars began to be transferred to the Mongolian nations, compare J. Eberh. Fischer, *Conjecturæ de gente et nomine Tatarorum* in his *Quæstiones Petropolitanæ*, p. 46, and his *Sibirische Geschichte*, T. I. p. 28, 142.

FIGURE 2.8. Johann Friedrich Blumenbach, *On the Natural Variety of Mankind* (1795), 269.

the moment he recognized the resemblance, the entire vast drama of human origins, wandering, and settlement unfurled before him. The Europeans, he suddenly realized, were the most direct and least degenerated descendants of that primal group, which must therefore be called "Caucasian." The other four races must then represent greater degrees of degeneration—hence the square heads of the Asiatics and the narrow heads of Negroes.

Everything was explained.

Dread

For many recent readers, Blumenbach has at this point said more than enough to establish his bad preeminence in the history of scientific racism. The indictment against him contains many counts.

- By recognizing divisions among humankind based on appearance and geography, he has laid the foundation not only for the study of race but also for racism.
- By describing Caucasians as the "primeval" race that has suffered the least change since the beginning of human history, he has implied that Caucasians were the norm or ideal, and that non-Caucasians were belated and "degenerate."
- By describing the Caucasian skull as the most beautiful of all, he has effectively ranked races based not on scientific evidence but on his preferred aesthetic.
- By acquiring skulls in the first place, he has implicated himself in the processes that produced skulls. His Georgian skull, for example, may have come from a woman taken captive by the Russians in a war against the Ottoman Empire and used as a sex slave until she died of venereal disease.[27]
- By ranking races based on his skull collection, he has anticipated some of the more ghoulish features of the Nazi genocide against Jews, gypsies, and others.

In all these ways, it is alleged, Blumenbach was a perhaps unwitting but still complicit pioneer in the ideology of white supremacy, his professed humanitarian commitments notwithstanding.

A substantial list of charges. But a good defense attorney could respond to all of them, arguing that:

- He is the precise opposite of a racist in that he insists that the human species is one and indivisible.

- He bases his varieties and his speculations about human origins on the physical evidence of skulls and, at least by allusion, on divine revelation.
- His "Caucasian" group includes not only Europeans but Asians up to fifteen hundred miles east of Ukraine as well as North Africans.
- His use of the Latin *degeneratio* has no negative connotations.
- The beauty of the Georgian people is widely attested.[28]
- He actually calls attention to his own cultural preferences (saying, "according to our opinion of symmetry"), and to the distribution of beauty and ugliness in all races.
- To describe Caucasians as beautiful is to attribute beauty to all humans prior to degeneration.
- He never undertook any measurements of skulls, explicitly denied that such measurements could have any value, and used skulls more as mile markers and illustrations than as representatives of distinct natural categories.[29]
- The skulls he acquired had been collected primarily from European sources and had been sent to him by other scientists and collectors.
- He would have been horrified by Nazi racial theories and genocidal practices; nor is there any evidence that Nazi ideologues cited Blumenbach, whose idea of a unified human species would have been offensive to them.[30]

These are strong, even on some points decisive responses, but the number and tone of the accusations suggest that something about Blumenbach, who was recognized in his time and for years thereafter as a powerful voice against racialist thinking and even against the concept of race itself, strikes many readers in recent years as darkly questionable. Why is Blumenbach so insistently accused of holding positions he rejected and blamed for crimes he did not commit? And why do his critics invariably operate in the mode of implication or insinuation?

This mode deserves some attention. Virtually none of Blumenbach's critics fails to credit him with good intentions; virtually all suggest that he was either indifferent to the real import of his views or unaware of the unconscious forces that led him to be sympathetic to arguments he repeatedly and explicitly opposed.

- One critic asserts, for example, that while Blumenbach was "one of the least racist thinkers," he was nonetheless also "*practically* a dictionary definition of a racist."[31]

— In his history of racism, George M. Frederickson includes Blumenbach, who, he says, "*opened the way* to a secular or scientific racism . . . *whatever his intentions*."[32]

— In an influential reading, Stephen Jay Gould notes that Blumenbach, who "*meant well*," insisted on the unity of mankind, campaigned against slavery, praised Negro authors, knew and corresponded with Negroes personally, and said that slaves were morally superior to their masters—yet "*nonetheless* . . . ended up with a system that placed a single race at the pinnacle." Since Blumenbach's "ideas *reverberate* through our wars, our conquests," Gould says, a line can be drawn from Göttingen to Auschwitz: "Where would Hitler have been without racism?"[33]

— Describing Blumenbach as "a crucial turning point in the history of white people," Nell Irvin Painter depicts him as a deeply conflicted figure whose unconstrained ambition and lack of self-awareness led him to substitute aesthetic preferences for scientific evidence, with the result that, bewitched by beauty and captive to his unexamined prejudices, he was *oblivious to the larger horrors of history* in which theories of race were linked to slavery.[34]

— Jacob Bronowski accuses Blumenbach (falsely) of using anatomical measurements to classify the families of mankind, and then, invoking the darkening skies over Europe in the 1930s, says that "Although there is no suggestion . . . that the skulls were to support a racist division of humanity. . . . *All the same*, from the time of Blumenbach's death in 1840 the collection was added to and added to and became a core of racist, pan-Germanic theory."[35]

— In a history of *The Rise and Fall of the Caucasian Race*, Bruce Baum writes that Blumenbach "*implicitly suggested*" that races were categories of nature.[36]

— Emmanuel Chukwudi Eze, editor of an anthology of writings on race and the Enlightenment, makes the case more directly, saying that, "*although he believed himself* opposed to racism," Blumenbach's representation of the relation between Caucasians and others demonstrates that he was mistaken in his belief.[37]

— In the 2020 student demonstrations at Göttingen University, Blumenbach was accused of having "*laid the foundations* for a branch of anthropology" that justified the oppression of "non-white peoples" (*nicht-weißer Menschen*).[38]

The implication seems to be that Blumenbach should not merely have led his era in humanitarian and progressive sentiments but should have taken a standing leap out of his own milieu, proleptically refuting some positions and embracing others that would arise long after his death. This is of course an unfair demand to place on anyone, which may be why it is only implied. But since none of Blumenbach's critics has a single suggestion to make about positions he ought to have taken, we might ask, What should he have said?

Should he have denied variation of any kind? That would be to ignore common sense and fail to recognize particularity. Should he not have mentioned the standard of beauty? Why not, since he acknowledged the bias involved? Should he have confined himself to books and refused to consider skulls or other physical evidence at all? That would be odd for a natural scientist. Should he have declined to make any inferences about the history of the human species? Why, if he thought he had understood facts that had previously been accounted for only by myths or fables? Should he have designated some other variety as the primeval one? On what basis?

Operating in the mode of insinuation, Blumenbach's critics suggest that while he may have been personally innocent of all charges and even exemplary in his values and character, the difference between him and absolute evil was not itself absolute, with the one passing insensibly into the other. Since his stated values and commitments are identical to those of his critics, however, the possibility arises that they are troubled by the suspicion that their own views, like his, might be secretly compromised in ways they do not recognize, and that a probing self-examination might reveal that they too are somehow tangled up in a tragic world of hatred and violence that they have inherited and from which they cannot quite get clear.

There is a Danish word for the effect Blumenbach creates in some recent readers: *angest*, or in German, *Angst*, a word that entered the philosophical lexicon in 1844 with the publication of Søren Kierkegaard's *The Concept of Dread* (1844). Kierkegaard's subject in this book is "hereditary sin," which he explores through a meditation on the Biblical account of the fall of man, a simple event that harbors, in his reading, extraordinary depth and complexity.[39]

The question is how mankind fell into sin, and when—at what point precisely. This seems like it should be an easy question until one tries to locate the exact moment of the transition from innocence to guilt. On closer textual inspection, the one seems to run insensibly into the other so that you cannot mark out the limits between them. Adam's eating of the apple? Clearly a culpable act, but almost redundant, more like a seconding of Eve's motion than a hinge in the history of mankind. Eve's eating of the apple? Definitely, but before she ate, she had to desire to eat. But before she desired, she had to hear the question, "Did God say thou shalt not . . . ?" and thereby entertain the possibility of judging the word of God. Indeed, her very participation in the conversation may constitute an act of disobedience, a possibility Kierkegaard alludes to when he says that it is "language itself" that addresses Eve, and that in understanding the serpent's question she is already lost.

In fact, however, she is born lost. As Kierkegaard says, the serpent succeeds by appealing to transgressive inclinations deposited in her like a primeval germ as a result of her creation as a secondary or "derived" being, doubly distanced from God and from Adam. Her innocence is no defense; God is not "innocent." Only flawed and derived beings are innocent; indeed, their innocence is the mark of their susceptibility to guilt, for which dread is the proper name. A state of "*sympathetic antipathy* and *antipathetic sympathy*," dread stands at—or rather just beyond, it is difficult to tell—the last extremity of innocence.[40]

Angst-ridden people suffering from the sense that they are laboring under the burden of some inherited sin turn to the past in search of the moment of transgression when that burden was incurred. If only they could find that moment, they feel, they could think their way back into a condition of innocence and resolve to avoid the temptation to sin. They discover, however, that their fate was determined even before the moment of decision by some undetected and even unsuspected prior weakness or vulnerability of which they had been unaware. What hope then? Almost none—but if there is any at all, it will depend on an extreme vigilance, constant suspicion, and a zeal to condemn lapses.

For modern Western antiracists, Blumenbach is Eve. His work is studied for signs of the corruption he did not suspect and of which he thought himself entirely innocent. His richly appreciative sense of nature's immense variety, his generous admiration of beauty in whatever form it might take, his creation of categories that might prove "serviceable to the memory," and his tireless pursuit of the formative force—all apparently honorable, all quite brilliant in their conception and formulation, and yet all somehow disastrously wrong, all sliding helplessly into oppression, imperial conquest, race-hatred, rape, venereal disease, slavery, sex trafficking, grave robbing. . . .

It is of course as absurd to compare Blumenbach to Eve as it is to compare him to Goebbels. But Eve, in Kierkegaard's reading, is a kind of synecdoche for the impossibility of drawing a clean line between before and after, innocence and guilt; and Blumenbach founded his entire theory—the theory that "made the modern human intellect what it is"—on the principle that in the domain of the human, everything is interrelated, and not even the difference between black and white is absolute.

Falling

For some reason, blame for the fall of man always lands on Adam and/or Eve, while the serpent escapes all accountability. They made choices and took actions that changed everything, while the character of the serpent was fixed and unalterable—it had always been that way, nothing better could be expected from it, and it would be pointless to condemn it.

Does Blumenbach have a serpent?

Five years older than Blumenbach, Christoph Meiners began his career as a professor at Göttingen at the same time Blumenbach entered the university as a graduate student. They came from the same milieu, lived in the same small town, worked in the same small university, belonged to the same small academy, traveled little, and read many of the same books. Like everyone, Meiners was aware of his brilliant younger colleague and especially of his success, and undertook a massive project, *Grundriss der Geschichte der Menschheit* [Outline of the history of mankind] (1785), in an attempt to create a "new science" of human existence that could compare and compete with Blumenbach's concept of degeneration. From a certain distance, the two men might appear as like as one egg to another, sharing the same background, training, interests, institutional setting, and ambitions.

But there were differences. Meiners was a speculative philosopher, a pontificator, an ideologue, and eclectic to the point of dilettantism. Personally, he was contentious, vain, irritable, arrogant, and dogmatic. Intellectually, he was dogged but unoriginal, incapable of abstraction. Politically, he was reactionary. Socially, he was contemptuous of those outside his group of middle-class northern Germans. Professionally, he was marginal, the author of unadmired books on various topics who eventually became a university administrator. In all respects, Blumenbach was the opposite.

The intellectual differences between the two men converge on the issue of human variation, on which they held differing views.

— Blumenbach noted that the skull of the Georgian female was beautiful "according to our opinion of symmetry," while Meiners

FIGURE 2.9. Christoph Meiners. Public domain. Source: Wikimedia Commons.

bluntly declared non-Europeans ugly, making no allowance for a European bias.

- Blumenbach attributed an original beauty to the entire human species; Meiners saw beauty, along with intelligence and virtue, only in Caucasians, especially those living in northern Europe.[41]
- Blumenbach thought all peoples could, under the proper conditions, rise to any level; Meiners disagreed.
- Blumenbach did not regard differences as unalterable; Meiners did.
- Blumenbach ultimately rejected geography as a category for the analysis of peoples, and regarded all national differences as inessential; Meiners said he wanted to "discover peoples through lands, and lands through peoples."[42]
- Blumenbach used the Latin *degeneratio* to mean deviation or transformation; Meiners used the German *Degeneration* to mean deterioration.
- Blumenbach insisted on the use of physical evidence such as skulls; Meiners rejected such evidence, confining his study to written records including travelers' reports and histories.

— Blumenbach was universally admired for his learning and equable temperament; Meiners was mocked by his contemporaries as "Mongol Meiners" (see below) and criticized for sloppy methods and dogmatism.[43]

Beyond these differences of opinion, capacity, understanding, and stature, the two men parted ways on the fundamental question of whether the category of the human contained one species or more than one. Blumenbach was committed to the former position, which was consistent with faith in the dominant understanding of the Biblical account of the origins of humankind in a single primal pair. Meiners was a polygenist for whom the Biblical story was a childish and misleading myth. In his account, "two primary lineal stems"—the European and the non-European—were responsible for the world's peoples; on reconsidering the matter late in life, he declared that there might be dozens of origins, each giving rise to a separate species.[44] The "human species," Meiners said in his *Outline*,

> appears to originate from two main divisions, the Tartar or Caucasian, and the Mongol; the latter being not only considerably weaker in body and mind, but also more lacking in virtue than the Caucasian division; for its part, the Caucasian division is divided into two races [*Racen*], the Celtic and the Slavic, among which the former appears to be more intelligent and richer in virtue. . . . European nations are superior, even in a state of savagery and barbarism, to savages and barbarians from other continents, in their sense of virtue, in their receptivity toward Enlightenment, in their constitution, in their laws, their manner of warring, their comportment toward women, slaves, and vanquished enemies.[45]

Not content with this disposition of talents and abilities, Meiners, who was as committed to division as Blumenbach was to unity, went on to add that the Celts themselves had a virtuous, beautiful, and intelligent core consisting of the Protestant Teutons of northern Germany, who were in all essentials far superior to the Catholics in Ireland and the south of Germany.

Meiners held all non-Teutons in degrees of contempt, but he was particularly scornful of indigenous Americans and especially Negroes, who were, in his eyes, human only in a restricted sense. The following passage from a 1790 essay called "Of the Varieties and Deviate Forms of Negroes" gives an impression of the energy Meiners could command when aroused.

> The ugly, stupid, and unruly Negroes certainly make up the greatest number of the original inhabitants of Africa. . . . Among all slave traders and planters, it is accepted as a settled article of experience that the Negroes on the east coast are uglier, stupider, and more savage than those on the coast of Guinea and, furthermore, that ugliness, stupidity, and indomitability generally increases the deeper Negroes live in the interior of Africa or have been brought out from the interior. The Negroes from innermost Africa are almost without exception cannibals. They have a horrible tiger-like, hardly human look and pointed or jagged teeth that close together like pinking shears, or the teeth of foxes. Many of them are so unruly and desirous of human flesh that they bite off large pieces of flesh from the arms or legs of their neighbors and fellow slaves and gulp them down. Slave traders from ships either do not buy such cannibals at all or they buy them at a much lower price than other Negroes. One specimen from the remote nations comprised of creatures that Römer said can hardly be called humans was so mistrusted by European buyers that he was sold from one ship to another for half a year despite the fact that the other Negroes had broken off his teeth, as if he were a beast of prey. When a certain captain had the misfortune of getting this monster among other slaves for the second time, he had some cannon balls tied to his neck and had him thrown overboard.[46]

Meiners defended slavery, which, he said, was not as burdensome to the slaves as it might appear inasmuch as the remarkably high pain threshold of Negroes suited them to the rigors of subjugation. As the example of the cannibal "specimen" indicates, however, he believed that only the better class of Negroes should be enslaved, for only these could be trusted not to eat their fellow slaves; indeed, they might even be expected to show an appropriate gratitude for having the opportunity to observe and benefit from the civilized practices of Europeans. "All credible writers," Meiners noted, "testify unanimously that the Negroes are improved by their removal to the Sugar Islands, or to the American mainland, and that the Creole Negroes, or the Negroes born outside Africa, are less stupid and less unruly, and, for that very reason, much more useful than the original Africans."[47]

In this humanitarian vein, Meiners pointed to the direct benefits accruing to slaves from the occasional interbreeding with whites, a plantation happenstance that could, he said, "ennoble the blood" of the lower race, thereby improving them in as few as three generations. So manifest were the benefits of slavery that the abolition for which some zealots were arguing would be harmful to everyone, even to Africans, who should, if they thought about the matter at all, consider themselves fortunate to have escaped the degrading forms of servitude and subjugation prevalent in their home countries and pleased by the elevation of their stock as a consequence of miscegenation. For Meiners, this elevation argued for slavery as a form of philanthropy rather than against it as a form of cruelty and exploitation.

With such performances, it is not difficult to understand why the high-minded intelligentsia of the late eighteenth century did not take Meiners to be a serious scientific mind. There are, as one scholar has noted, "only a handful of 19th century articles that address Meiners, and they tend to dismiss him as a second-rate scholar. . . . It appears that Meiners' penchant for lashing out against those who disagreed with him—and that was almost everyone—tarnished his memory for decades after his death. Nowadays, his cold-hearted bigotry is an embarrassment to modern students of cultural anthropology, sociology, and psychology, three fields in which he was unquestionably a pioneer." In this respect, Blumenbach was ahead of his time; references and comments in his published work suggest that he "discreetly but profoundly despised" his colleague.[48]

And yet it is possible to make a case for Meiners's virtues. The historian of anthropology Britta Rupp-Eisenreich has made the effort, arguing that Meiners was a fearless opponent of prejudice and received wisdom, that he undermined the notion that nations or cultures had a singular claim to virtue, expanded the horizons of thought by treating all peoples as worthy of study, and sought to improve the morals of his civilization by his work. All admirable, she says, considered in isolation; it is only "the thrust of his argument" that people have found objectionable.[49]

A stronger case might be made for Meiners's influence. It was Meiners not Blumenbach who appears to have influenced Virey, the French racial theorist whose *Histoire naturelle du genre humain* (1800; translated as *Natural History of the Negro Race*) divided mankind into two distinct species, the fair and white and the ugly and black. Virey focused particularly on the physical, moral, aesthetic, and intellectual deficiencies of Negroes to which Meiners had drawn such memorable attention, suggesting that Negroes were so utterly different from other groups that they appeared to represent "not only a *race*, but truly a *species*, distinct from all other races of men known on the globe."[50] And it was Meiners not Blumenbach who was cited as an authority in the British Parliament during debates on the slave trade in 1788 (on the anti-abolitionist side).[51] And Meiners not Blumenbach who influenced the Scottish race supremacist Robert Knox along with an entire generation of like-minded people in France and Italy. And Meiners not Blumenbach who was mentioned by Gobineau in the first volume of his immense tract *An Essay on the Inequality of the Human Races*.[52] And it was Meiners and certainly not Blumenbach who was described by a pro-Nazi German anthropologist as the "founder of raciology."[53] Disrespected when he was not ignored in his time, Meiners eventually found his way into the mainstream.

The argument has even been made that Meiners's influence extended to Blumenbach himself.[54] Absurd on the face of it, this suggestion has some evidence behind it. It was, for one thing, Meiners who coined the racially charged term Caucasian, with Blumenbach adopting the term in the third edition of *Natural Variety*. For another, Blumenbach's insistence on the "beauty" of the first humans may well have been a response to Meiners's judgment that all non-Europeans were ugly. And for another, while Blumenbach initially resisted the term Meiners preferred—race, or in German *Racen* or *Rasse*, which suggest "breed"—he eventually conceded the point, perhaps out of concern for the possibility that his own preferred terms *Varietät* and *Spielart* carried the implication of superficiality and contingency, whereas race suggested a biological reality that could be studied by a scientific method.[55] Whatever the reason, race became Blumenbach's preferred term for subgroups of species. In *A Manual of the Elements of Natural History* (1779), he spoke of his five groups as "*Hauptrassen*," or primary races, and treated race and variety as synonyms. And in 1794, a year before the third edition of *Natural Variety*, he wrote of the "five races of the human species."[56]

So while Blumenbach may have disapproved of his vulgar and embittered colleague, and may have been revolted by his opinions, he did not repudiate him altogether, and even seems to have engaged in a kind of dialogue with him, thereby inviting insinuations of sympathetic antipathy or antipathetic sympathy, and enduring, posthumously, the humiliation of being identified as the secret source and unwitting exponent of opinions and judgments Meiners held proudly, in full view of the public.

We might think of the efforts to make Blumenbach into Meiners as something like a translation error, similar to that represented by the translation of *degeneratio* into the German or English degeneration, or *Varietät* into *Rasse*. But—Blumenbach's critics might ask, in their insinuating way—are these really mistranslations? Blumenbach may have insisted that by *degeneratio*, he meant only deviation, but if he had been writing in German he would in all likelihood have used *Degeneration*, with its connotation of deterioration. Blumenbach's fluency in Latin notwithstanding, *Natural Variety* is what might be called an "original translation," a text conceived in one language and expressed in another, so such questions arise about the entirety. In fact, in the German original of *Contributions to Natural History*, Blumenbach gave the section in which he explained the concept of *degeneratio* the title of "*Die Ausartung der organizierten Körper*" (The degeneration of organized bodies), and while *Ausartung* had, in Blumenbach's time, a neutral tonality, the word must have had a certain lining or shadow, a susceptibility to inflection, for since that time the word has acquired connotations of excessiveness or extremity that align it with the modern meaning of degeneration.[57]

The most serious issues regarding translation focus, however, on the 1865 English translation I have been citing. The book by which Blumenbach became known to large numbers of English-speaking readers was produced under the sponsorship of the Anthropological Society of London (ASL), an organization founded in 1863 by James Hunt, a wealthy speech therapist and devoted admirer of Meiners. Hunt was the author of *On the Negro's Place in Nature* (1863), the gist of which can be inferred from the title and from the fact that Hunt supported the Confederacy in the American Civil War. Under his leadership, the ASL was dedicated to polygenism and race ranking. An advanced thinker on the subject of evolution, Hunt believed that the races evolved out of separate ape species, so that Negroes and others could be considered as "distinct species from the European."[58]

Like the primeval germ aroused by the subtle odor of the male, Hunt sensed in Blumenbach a distinguished (unlike Meiners) colleague and ally whose work deserved a wider audience in the English-speaking world. He entrusted the task of translating Blumenbach's major works to the vice president at the ASL, a Cambridge barrister named Thomas Bendyshe. In a remarkable piece of inferential research, John S. Michael has determined that the translation produced under Bendyshe's name was actually done by two people, one of whom (who may have been Bendyshe himself) shared the polygenist views of Meiners and Hunt and the other of whom did not.[59]

Bendyshe was a man of clear but contradictory and often mistaken views, one of which was the belief that Blumenbach was, essentially, Meiners. Indeed, in translating Blumenbach's Latin into English, he effectively translated him into Meiners—transposing the major figure of Blumenbach, we might say, into a Meiners key. In the introduction to his translation, he declared that direct statements by Blumenbach to the contrary notwithstanding, Blumenbach believed in a plurality of races and did not think that all five of his groups should be placed "in the same rank" or given the same "zoological value." Bendyshe cited a French zoologist, Isidore Geoffroy Saint-Hilaire, to the effect that "the unequal importance of these races in a zoological point of view, is also, at least by implication, admitted by Blumenbach."[60] Which of course it was not.

Exercising the vast but concealed power of the translator, Bendyshe produced a Blumenbach that conformed to his characterization. Michael cites a number of instances where non-Caucasian varieties are described in the translation, but not in the original text, as ugly, unpleasant, or unattractive. Moreover, Bendyshe interpolates judgments where none exist in the original, describing the Mongolian nose as "apish" where Blumenbach had characterized it as "small and flat," and black skin as "repulsive" when Blumenbach had made no such statement.

The descriptions of Caucasians are distorted in the opposite direction by Bendyshe's insistent nudging of Blumenbach's variously nuanced Latin terms into the register of the aesthetic, creating the impression that Blumenbach was more a Romantic sensibility ravished by sensual impressions than an Enlightenment scientist making a good-faith effort to be precise. Perhaps the greatest distortion of all, however, is one that, given Blumenbach's later adoption of the term race, might seem unremarkable, the translation of Blumenbach's Latin terms *varietas*, *stirps*, *stemma*, and *gens* as race. The "racial face" discussed earlier is Bendyshe's term; Blumenbach speaks of *facies gentilitiae*, which might be more accurately translated as "tribal" or "national" face.

In short, while presented as an instance of *degeneratio*, Bendyshe's translation is actually a degenerate version of Blumenbach's text—a deviation downward.

In recent years it has been North American academics with progressive views who have been depicting Blumenbach as the secret fellow traveler and enabler of Meiners, and the hidden inspiration of those who followed in his line. They have been able to press this point in large part because they all rely on Bendyshe. It is difficult to decide who comes off worse in Michael's investigation, Hunt and Bendyshe or more recent commentators such as Stephen Jay Gould, Robert Wald Sussman, E. C. Eze, and others who, lacking Latin, German, or in some cases scholarly integrity, take Bendyshe for Blumenbach, making no mention of Meiners. The consequence of their inattention is that while Blumenbach (along with Hume, Kant, Voltaire, Goethe, and others) has been darkly implicated in the history of racism, Meiners, who contributed nothing to enlightenment in either a general or a specific sense, and who actually influenced subsequent racist thinking, has been allowed to slither away unnoticed.[61]

So different are Blumenbach and Meiners that their proximity at Göttingen must at one time have seemed a freak of contingency, almost as if different species had been placed in the same enclosure. To the extent that it considers Meiners at all, the recent consensus is, however, that the relationship was anything but accidental, and that there was far more intellectual compatibility between them than met the eye. Colleagues in "raciology," which might mean either the science of human variation or the study of racial inequalities, Blumenbach and Meiners are now considered by many to be mere variants.

We may still ask about the nature of the variation. Is Meiners the degenerate form of Blumenbach's original perfection, a sad falling off from Blumenbach's pristine theorizing, or is he a rude and unevolved version of the more sophisticated and humane Blumenbach? Which is the skull and which the misleading face? Is Blumenbach the natural and foundational frame and Meiners the careworn soft tissue beset by gnats and other worldly stressors, or is Meiners the changeless structure of hatred and division that can be masked but never fully negated by the kind and loving face of universal philanthropy? What is the real difference between the two?

It is tempting to posit a clean or "species" distinction between them on the basis of distinction itself, a measure according to which Blumenbach excels and Meiners scarcely registers. But by what objective measure can we say that one is more distinguished than the other? Blumenbach may have named five primary races, but anthropologists today do not endorse that division, which has found a sanctuary only in the US Census. Blumenbach may have invented physical anthropology, comparative anatomy, and the modern human intellect, but those would have been invented without him, and the first has been largely discredited or replaced by other terms. Indeed, Blumenbach's suggestion that skulls should be the source of evidence about human variation was to mislead anthropologists for over a century before being dismissed with prejudice. Nor does anyone still speak of the primeval germ, the formative force, or degeneration. Nor are the Caucasus Mountains still considered the ancestral home of humankind. Nor do scientists continue to credit the aerial origin of the Ethiopians' pitchy blackness. And Blumenbach's dogmatic insistence on a single undivided species can seem to our diversity-obsessed age blind to history, inattentive to significant distinctions, even vague and sentimental—a pretentious form of "We Are the World" or United Colors of Benetton.

In fact, none of Blumenbach's ideas enjoys any currency today in the forms he gave them. By contrast, the feeling to which Meiners gave early and uncompromising voice—that humanity is divided into categories, and that our group (whoever we are) is beautiful, virtuous, and smart, while the others are to varying degrees ugly, wicked, and stupid—enjoys nearly universal support, and often passes for common sense, which is surely the highest praise that can be given to an intellectual, especially one who, like Meiners, prided himself on being a contrarian. Considering the frequency with which this feeling is associated with the concept of race that Meiners promoted and the Caucasian superiority to which he was committed, some might even argue that Meiners has a greater claim than Blumenbach to having "made the modern human intellect what it is."

The debate between Blumenbach and Meiners was restaged countless times over the century that followed, as intellectuals, scientists, and polemicists tried to work out whether humankind consisted of variations, with all categories understood as arbitrary fictions in a unified field, or of distinct species with deep and durable differences. Accommodating both meanings, the dialectical concept of race held these opposed views in suspension, which is why it was able to serve as the focus and flash point of confrontation on these questions, and also why so many debates about race got stuck on the definition of the word.

It is interesting to think how different the world would have been if the irreducibly ambiguous concept of race, with its biological and zoological implications, had never been adopted for the purposes of understanding human variation. What if Blumenbach, especially the early Blumenbach, had prevailed and Meiners (et al.) had sunk into, or had never risen from, obscurity? What if all humans, whatever their current conditions or manifest differences, had always been thought to be equal in fundamental dignity, rights, or potential, and equally distinct from primates? What if people had understood themselves to be living in a sfumato world of adjacencies, clines, and gradations, with forms running insensibly into each other so you could not mark out the limits between them?

CHAPTER THREE

More or Less Human?

MORTON / NOTT / DOUGLASS

The Value of Colored Life

So great was the disorder and so high were the stakes that a distinguished physician felt compelled to address the situation. Insurance companies insensitive to the realities of life in the South were losing money on their life insurance policies. Josiah C. Nott of Mobile, Alabama, a graduate of the University of Pennsylvania Medical School who had studied clinical medicine in Paris, took time off from his numerous and pressing responsibilities to offer his reflections "on the value of life among the colored population, which is becoming a very important consideration."[1]

Appearing as the lead article in the November 1847 edition of *The Commercial Review of the South and West*, Nott's article, "Statistics of Southern Slave Population with Special Reference to Life Insurance," stood out among the other articles in the journal, which concerned river and harbor improvements, the cultivation of sugar, the commercial importance of American lakes, American native wines, and the natural resources of Texas. The author was no mere actuarial or compiler of statistics but a man of wide experience and considerable public reputation, admired for his aristocratic manners and celebrated, even revered, for his discoveries about the pathology of yellow fever and his expertise in the use of leeches. A physician, an eloquent spokesman for the culture of the American South, a recognized scientific authority on human variation, a historian with a deep understanding of distant lands and cultures, and even something of a philosopher, Nott was capable of comprehending in a single capacious vision the practical considerations peculiar to the insurance industry and issues of a quite different kind that were beginning to occupy the pages of scientific journals in Europe and the United States.[2]

Nott could speak with confidence on a range of issues in part because he enjoyed the acquaintance of many leading scientific figures including the universally respected Philadelphia naturalist Samuel George Morton. The author of the landmark *Crania Americana*, Morton had dramatically reshaped the international discussion of human variation by connecting traits that many had considered the products of climate, long isolation, and deep-seated cultural habits to empirical data about skull size and shape.[3] While limited in its scope, Morton's work promised exciting discoveries yet to come as craniology matured into a discipline that could yield scientifically sound information about human groups and their inbuilt capabilities.

Conversant with Morton and other leading scientists, Nott was clearly addressing a more sophisticated audience than the subscribers to *The Commercial Review*. Insurance was but the innocuous flag under which marched a formidable belligerent force of argumentation whose true objective was larger and different in kind than the setting of rates. As a scientist and slave owner, the intensely ambitious Nott believed that he had a deeper understanding of the theoretical and practical issues surrounding the subject than virtually anyone else in the world. By training and experience, he believed, he was in possession of facts not generally known, facts that properly considered would not only place the insurance industry on a more rational and sustainable footing but would have a decisive impact on human self-understanding.

But first, the rates.

Insurance always involved risk, but the peculiar risks involved in insuring slaves had escaped attention, with the result that the industry was being conducted "pell-mell, without knowing anything of the probabilities." This ignorance produced needless suffering not only for the joint stock companies formed by "heavy capitalists" but for the "poor, honest, industrious and I may add unsuspicious man, who at the end of the year scrapes together a few of his hard-earned dollars to invest in a mutual insurance company, with the hopes of saving his wife and children from beggary when he is no longer able to toil for them." When slaves died before they had been expected to and the insurance company had to pay out, the man and his family lost their money—needlessly, Nott argued. The poverty of statistical information about the mortality of slaves was, Nott said, "positively disreputable in this enlightened epoch." Writing in a spirit of humanitarianism, philanthropy, and compassion for the yeoman

FIGURE 3.1. Josiah C. Nott. Public domain. Source: Wikimedia Commons.

investor, Nott undertook to repair the reputation of the epoch in this one particular.

The question whose answer would begin the vast process of correction was why so many slaves died before the actuarial tables suggested they would. It was not an easy question, for many facts suggested that slaves should live long and healthy lives. Negroes in the American South were provided with food and shelter in a temperate climate where their needs were attended to by a master who, Nott noted, generally treated them with "kindness and indulgence." Moreover, as Nott had determined in his medical research, Negroes seemed mysteriously exempt from various diseases including cholera, malaria, and yellow fever.[4] In many respects, Nott said, Negro slaves were a privileged group, especially compared to their African relations who were exposed to the elements and the animals, and to enslavement by "barbarian chiefs." They were even better off than Negroes living in the North, who enjoyed "the abstract delights of liberty" but suffered the concrete miseries of cold weather against which they were by nature powerless to protect themselves. The Southern Negro was at a racial apogee, having attained "his greatest perfection, physical and moral, and also his greatest longevity, in a state of slavery." It would seem, in short, that the slaves of the South would be an eminently insurable population.

An advisor who merely accepted these facts would, however, provide false and misleading information to his clients. Slaves die in numbers greater than might be expected for several reasons not accounted for in the above résumé. For one thing, the insured slave presented many temptations that could not always be resisted. In the case of an "unsound Negro" worth less to his owner than the premium, the underwriters could not expect fair play. An "unfeeling master" might come to regard the slave "in the light of a superannuated horse" and simply let him expire in order to collect the insurance. The sadly universal truth was that "any man who will drive a horse cruelly will drive a negro or operative to death, if he can gain anything by so doing." Indeed, unsound Negroes were sometimes insured fraudulently on the certainty of an early demise. Such things, Nott wrote, were to be expected because "human nature is the same everywhere, and at all times," and the booming voice of the "Almighty Dollar" easily silences "the soft, small voice of humanity."

Considering all this, the counterintuitive conclusion virtually announces itself: "It would be unsafe to insure negroes on plantations in the country."

An even greater insurance risk, and a far more compelling theoretical issue, concerned the growing mulatto population, not all of whom were enslaved. Having lived in South Carolina and Alabama, and having traveled frequently to New Orleans, Nott had ample opportunity to make a professional observation of this population. He noted that mulattoes tended to be "intermediate in intelligence between the blacks and whites," "less capable of enduring fatigue, exposure, and hardships of all kinds, than either blacks or whites," and more disease-prone than "pure blacks." Female mulattoes were particularly fragile and were "bad breeders and bad nurses" to boot. Many had difficulty conceiving at all, especially when paired with another mulatto.

The relative infertility of the mulatto had been remarked on by others, but Nott, with his physician's keen eye and his theorist's sensitivity to patterns and regularities, noticed something others had missed, a distinction between kinds of mulattoes. Some mulattoes—mixed-race "creoles" of the kind found in great abundance in Charleston, Mobile, and New Orleans—flourished. Robust, attractive, and prolific, they seemed almost to constitute "a *new*, or *distinct race*." Others, formed of the union of Negroes and Teutonic or Celtic whites, on the other hand, languished, a fact Nott attributed to the greater degree of difference between the races of the parents. Short-lived, weak, infertile, and mentally limited, this group represented an unnatural commingling of superior and inferior elements. They seemed to belong not to multiple races but to no race at all; they were merely the corruption of the parent stocks. Mulattoes of this kind should, he proposed, be considered degenerate "hybrids," human mules that could be produced but could not reproduce and were doomed to work out their own extinction.

This insight became the germ of a deeply suggestive theory of human variation, the first to appear on American soil. Human beings of different groups, Nott ventured, had "certain affinities or repulsions" that were registered not in their unlicensed animal inclinations alone but more directly in their ability to procreate. The French, Italian, Spanish, and other darker-skinned whites had a greater biological proximity or "affinity of blood" to Negroes, and their unions enjoyed what amounted to the grudging permission of nature in the form of offspring.[5] The fairer Caucasians, by contrast, stood at a greater biological distance from Negroes, and nature, insisting on its own categories, rebelled at attempts to reconcile or commingle the two, effectively declaring that enough was enough.

The conclusion to be drawn was that while human history might represent a vast chronicle of promiscuity and confusion, races were in essence fixed and unalterable, and transgressions against the order of nature were punished with feebleness and infertility in the offspring. The compromised biology of Caucasian-Negro progeny proved the existence of races—natural categories of humans, primordial original types. Races possessed a greater integrity than some of their members and policed their boundaries even when individuals did not. The Teutonic-Celtic Caucasian type was particularly and appropriately vigilant in this respect.

Nott had written to the distinguished Morton in 1845 about these affinities and repulsions, and Morton, knowing a superior mind when he saw one, incorporated Nott's work in an important essay two years later, "Hybridity in Animals and Plants," in which he proposed a "theory of *repugnance*" that held that although all humans had a latent power of hybridity and even a disturbing propensity toward illicit mixtures, those races with the greatest degree of difference found the bare prospect of mingling their blood with certain other races morally offensive and physically loathsome.[6] Human feelings in such instances were evidence in the organism of the laws of nature: revulsion at interracial intercourse testified to the fundamental organizational clarity of nature, a clarity denied

FIGURE 3.2. Samuel George Morton. Public domain. Source: Wikimedia Commons.

by sentimental humanitarians prating on about the unity of the human race. Morton's examples, like Nott's, were Negroes and Caucasians—the extremes of the racial spectrum, the only remnants of the aboriginal racial groups, the only pure tones left from the mighty chord of creation.

Happy to have the eminent Morton's agreement on key points, Nott still sometimes found Morton's judgment flawed, perhaps as a consequence of the restricted opportunities for direct observation in Philadelphia. What else could explain Morton's uninformed, misleading, and entirely gratuitous comment—the only one Morton ever made on the subject—that the natural antipathy between Caucasians and Negroes had been partially overcome by extended proximity, and "above all other means by the moral degradation consequent to the state of slavery" that had numbed nature and permitted miscegenation?[7]

Nott saw the matter differently. As a slave owner surrounded by an ever-increasing mulatto population that he could observe at close range, he felt he had a cleaner, sharper understanding of the matter than sermonizing Northerners. These misguided moralists often failed to see what was so clear to Nott and his Southern friends, that the institution of slavery, while challenging to administer, conceptually unappealing, and even perhaps reprehensible from a moral perspective, was not only necessary for the American economy but was actually based on a respect for nature and natural differences.

As evidence for these differences and for the permanence of what he would subsequently call racial types, Nott—still writing about insurance rates as they pertained to slaves and mulattoes—pointed to recent work on Egyptian monuments, paintings, and hieroglyphs that showed that the fabulous civilization of ancient Egypt was biracial, with the races as distinct then as they are now, and distinct in the same way. Historical research combined with close observation by fresh eyes had yielded the conclusion that the pharaohs, their families, and the ruling class in general were recognizably Caucasian, while the slaves and servants were plainly Negroes. With evidence from the dawn of civilization, Nott was able to declare that Caucasians were born to rule, while Negroes, being morally and intellectually limited as well as naturally mild and docile, were suited for servitude.[8]

Long experience in the South had proven to Nott's satisfaction that Negroes flourished in extreme heat; meanwhile, he noted, substantial evidence indicated that they were unable to function in the chilly North, with mulattoes faring even worse. Not only did the ingrained habits of all Negroes—indolence, imprudence, indifference to the future, inattention to the basics of self-preservation—expose them to the elements, but their very will to exist seemed to wither as the temperature dropped. This peculiar vulnerability, Nott said, was surely evidence that Negroes had been intended by God to live in their ancestral home of equatorial Africa, where Caucasians struggled to adapt. Modern science supported God in this particular. An article recently published in the *Boston Medical and Surgical Journal* asserted that mulattoes and free Negroes living in Boston were not as strong, healthy, or long-lived as white people living there.[9] Drawing on his own experience, Nott added that "even Cholera and Slavery combined, here, are far less destructive to the negro, than liberty and climate in Boston."

After submitting his essay on the value of life among the colored population to *The Commercial Review*, Nott reflected on the broader context of his reflections, and sent the journal an additional statement on "The Slave Question," which was printed beneath the article. Based on "*negro statistics*," this statement both anchored his conclusions in documented facts and unfurled his arguments so the reader could grasp the true dimensions of the issues he was addressing including the actual consequences of abolition, should such a thing ever happen. He invited his readers to contemplate the prospect of three million slaves—"indolent, improvident, and vicious as free negroes are everywhere"—turned loose by a federal proclamation of emancipation, and heading north. If this happened, Nott said, Northern abolitionists now piously clamoring for an end to slavery "would sooner vote that all the tribes of Africa should be turned over to the devil without benefit of clergy." The only other remotely viable post-emancipation plan was colonization; but while "it is much to be desired that some one would give us a project by which these millions of ignorant, stupid negroes can be successfully colonized, and kept from relapsing (as they are rapidly doing in St. Domingo [Haiti]) into African barbarism," nobody had come forward with such a plan. The horrifying implications of abolition constituted, Nott concluded, a question whose resolution we must trust to God, whose "hand at last . . . rules the destinies of nations."

Suffice it to say that none of the other articles in *The Commercial Review* took up any of the many subjects that Nott had found relevant to his argument. These included:

— the economics of the plantation
— the natural human inclination to profit even at the cost of the lives of others under one's care
— the different aptitudes, cognitive capacities, and moral characters of human races
— the differences between human races with respect to diseases and climate conditions
— hybridism in plants and humans
— degrees of human hybridism
— the "many peculiarities which belong to the mulatto"
— the differences in living conditions of people in the United States and Africa
— the moral and other arguments in defense of slavery
— mortality rates among black people in different climates
— the working conditions of slaves
— the political and other differences between the South and North
— the origin of humanity
— the source of human differences
— the status of the Bible as a source of knowledge
— the comparison of the skulls of ancient Teutons with those of contemporary Negroes

— the racial composition and culture of ancient Egypt
— the unalterability of racial characteristics
— prehistorical migratory patterns
— the effects of interbreeding on the "superior" and "inferior" races
— the reasons for civilizational collapse
— the likelihood of social disruption on an inconceivably vast scale if large numbers of Negroes moved north
— the practicability of transporting millions of people out of the country

What linked all these subjects and made them necessary and integral to insurance was the issue of race, which had both specific consequences for local business decisions and a huge nimbus of implication in which were floating a vast number of concepts, possibilities, and facts, all of them potentially, conceivably, or arguably connected. To think of race was, for Nott, to contemplate both the segmentation of nature and a universal but mysterious principle of connectedness, imperceptible to the uninformed eye, that linked times, places, bodies, nature, systems, categories, histories, nations, insurance companies, and, most mysterious but also potentially the most determinative of all, the will of God.[10]

No Dishonor to an Hottentot Female

A combative and even assaultive writer, Nott provoked and welcomed controversy. But his views on race and racial characteristics were not controversial, being entirely compatible with the common sense and experience of vast numbers of people in the United States and in Europe. On the subject of hybridity, Nott spoke with the authority of conviction and consensus, and was at most a few degrees crisper in his articulations than the esteemed Morton. What was at first perhaps only intuited by most of the readers of his wide-ranging pronouncements was an implication that others who lacked his scientific stature or American audience had been circling around for years, that Negroes were not merely different from white people in various observable and measurable ways but that they stood at the margin of the human in a category for which race was almost an inadequate word.

Nott had a better word. If the mulatto offspring of Caucasians and Negroes were to be considered in their true aspect as unnatural and degenerated hybrids, as the mortality statistics reported by the *Boston Medical and Surgical Journal* strongly suggested, then the two groups were more like species than races.

Morton had approached this clarifying recognition but had failed to state directly what his argument had plainly implied. Direct statement was Nott's strength, as he had demonstrated in his first foray into the world of anthropological theory, a brief 1843 essay on mulattoes where he argued that, considering everything—brain size, head shape, facial angle, nervous system, intellectual powers—Anglo-Saxons and Negroes are as unlike as swans and geese, or horses and asses. The comparison was clearest in the case of females, with the Caucasian woman's "rose and lily skin, Venus form, and well chisled features" standing in sharp and pleasing contrast to "the African wench, with her black and odorous skin, woolly head and animal features." In short, "*the Anglo-Saxon and Negro races are, according to common acceptation of the terms, distinct species, and . . . the offspring of the two is a Hybrid*."[11] The very title of this article was a model of argumentative boldness and clarity: "The Mulatto a Hybrid—probable extermination of the two races if the Whites and Blacks are allowed to intermarry."

Both the conventionality and originality of Nott's contribution can best be assessed by comparing it to the arguments of two others with whom he was in broad sympathy. In "The Mulatto a Hybrid," Nott had cited Edward Long, a colonial administrator in Jamaica and author of a definitive three-volume account of the island's history, geography, economy, governance, and social structure.[12] Like Nott, Long had first-hand knowledge of his subject inasmuch as his family had lived for over a century among the slaves who provided the labor that drove the island's economy. Having prospered in this circumstance, Long, again like Nott, defended slavery not only because it produced profits but also because it suited the character of both races and was indeed much to the credit of both. And, like Nott, Long considered himself not merely a professional in a certain field but a participant in the intellectual culture of the day.

A practical man—a planter, slave owner, and administrator—as well as a polymath with developed interests in science, history, philosophy, and theology, Long did not restrict himself to statistics and documents. Drawing on his wide reading, he could refer to Voltaire, Buffon, Linnaeus, Hume, and others, and could speak with assurance of the "system of man" or "scale of nature," versions of the medieval concept of the "chain of being" in which all God's creatures were ranged in a ranked order from the lowest to the highest.[13] In this context, he offered his reflections on

FIGURE 3.3. Edward Long. Source: National Portrait Gallery, Washington, DC.

the subject of human variation, focusing on the points of abutment and transition between levels.

Of particular interest to Long in this respect were reports of a trained orangutan who "sat at table, unfolded his napkin, wiped his lips, helped himself, and conveyed the victuals to his mouth with the spoon and fork [and] drank wine in moderation." Possessed of a mouth that "*does not speak*" and a brain that "*does not think*," this creature, Long said, is more a "*wild man*" than an animal, and "fills up the space between mankind and the ape," reaching upward as it were toward the lowest human.[14]

With what Long regarded as their gross but imperfect resemblance to their masters, Negroes fell into a nearby category, the lowest humans reaching downward to primates—reaching so far down, in fact, that they touched, or could touch, even intimately. "Ludicrous as the opinion may seem," he wrote, "I do not think that an oran-outang husband would be any dishonour to an Hottentot female." For this obvious reason, he said, "the planters do not want to be told, that their Negroes are human creatures." A man of science, he drew the taxonomic conclusion: "When we reflect on the nature of [Negroes], and their dissimilarity to the rest of mankind, must we not conclude, that they are a different species of the same *genus*?"[15]

The key to understanding both the needs of the plantation and the system of nature as revealed by modern science was, Long insisted, the principle of diversity, by which he meant the fundamental differences between adjacent kinds that might be masked by trivial similarities and resemblances.

> If we admit with Mr. Buffon, that with all this analogy of organization, the oran-outang's brain is a senseless *icon* of the human; that it is meer matter, unanimated with a thinking principle, in any, or at least in a very minute and imperfect degree, we must then infer the strongest conclusion to establish our belief of a natural diversity of the human intellect, in general, *ab origine*; an oran-outang, in this case, is a human being, *quoad* his form and organs; but of an inferior species, *quoad* his intellect; he has in form a much nearer resemblance to the Negroe race, than the latter bear to white men; the supposition then is well founded, that the brain, and intellectual organs, so far as they are dependent upon meer matter, though similar in texture and modification to those of other men, may in some of the Negroe race be so constituted, as *not to result to the same effects*; for we cannot but allow, that the Deity might, if it was his pleasure, diversify his works in this manner, and either withhold the *superior principle* entirely, or in part only, or infuse it into the different classes and races of human creatures, in such portions, as to form the same gradual climax towards perfection in this human system, which is so evidently designed in every other.[16]

Where Linnaeus drew the species line between humans and simians, Long drew it between Negroes and Caucasians, two groups so deeply different that they could not possibly have had a common ancestor.

All this accorded with Nott's view of the matter. (And not his alone. Much of Long had already been echoed by another slave owner, Thomas Jefferson, whose 1785 *Notes on the State of Virginia* had voiced tentative doubts about whether Negroes and whites were of the same species, even mentioning "the preference of the Oranootan for the black women over those of his own species.")[17] But with even greater experience in a racially mixed world than Nott, Long had come to an altogether different conclusion on an important point. A hard man living in a hard world, Long had been witness to the disgustingly common practice of excessive familiarity between the races. "Many are the men," he wrote, "of every rank, quality, and degree here, who would much rather riot in these goatish embraces, than share the pure and lawful bliss derived from matrimonial, mutual love."[18] Nott could only agree that goatish embraces threatened not only the order and stability of the plantation but the natural order that justified it and the integrity of the race that owned and operated it. But he had to reject Long's next point about the viability of the mulatto.

Acknowledging that mulattoes are "actually of the mule-kind," incapable of reproducing among themselves, Long noted that they were not condemned to self-extinction but could survive and flourish if they interbred with whites, in which case the offspring could become in the course of a few generations almost white again.[19] Nott would have found the prospect that the descendants of mulattoes might erase all traces of transgression and regain an original whiteness offensive and scientifically counterproductive because it argued against the claim he and Long were making about the two races being different species. Indeed, if he had pondered the matter, Nott might have realized the head-spinning, stomach-churning implication Long had created, that given the realities of plantation culture, a white person a few generations hence might trace his or her parentage back to both the orangutan folding his napkin and wiping his lips and the host—or hostess—of the dinner.

Another writer who spoke directly to Nott's interests was Julien-Joseph Virey, a French polymath whose *Histoire Naturelle du genre humaine* (Natural history of the human species) appeared in 1800 when the author was twenty-five. Virey produced a revised and greatly enlarged three-volume edition twenty-four years later; this text was pillaged with great freedom by a South Carolinian, J. H. Guenebault, who "extracted" the part of Virey's text that dealt with Negroes and published it in 1837 in Charleston, which had at that time one of the largest slave markets in America.[20] Nott, who had just left Charleston for Mobile, Alabama when this work appeared, cited it in "The Mulatto a Hybrid" as the work of "a distinguished French naturalist."[21]

Nott was favorably impressed by Virey's prescient grasp of the importance for the study of race of skeletal structure and especially skulls, whose pertinence he had promoted with more energy than either Petrus Camper or Blumenbach. With only a short step from the skull to the brain, it was, Virey argued, possible to draw from statistics about skulls inferences about the intellectual capacity, cultural habits, and moral dispositions of

FIGURE 3.4. Julien-Joseph Virey. Public domain. Source: Wikimedia Commons.

the races. On that basis, one could claim—and Virey did claim—that these were as rigid and inflexible as the skull itself.

More than half a century before Darwin, Virey explained these differences in naturalistic terms as the consequence of an almost inconceivably protracted process of species self-modification from primates to humans. "Europeans"—his preferred term—had managed to complete the process, achieving full civilization while others had languished, remaining in their unimproved original condition. Black people in particular remained "in a stupid and brutish state; whilst the other nations on earth have approached more or less to social perfection." As to the present state of play, Virey was clear and specific. "Generally speaking," he wrote,

> the European is the most intelligent and polished of all the inhabitants of the Globe, after him are to be ranked the Hindou, the Asiatic of the Mongul race, as the Chinese, Japanese &c., who themselves

> are followed in the gradation of intellect by the Malayan tribes, the Calmucks and native American hordes; finally going down we reach lst, the Laplander, 2d the Negro, 3d the Hottentot and even the Cretin, 4th the Makaia, 5th the Negro Eboé, who . . . has a mouth shaped exactly like that of a Baboon. We pass then by imperceptible degrees from man to monkey. . . . [Indeed,] *the monkeys of Africa are more stupid and ferocious than those of Asia.*

The degrees and gradations may be nearly imperceptible as we slide down the chain of being, but for Virey as for Long, the extremes of Europeans and Africans were so different that the two groups had to be considered distinct species, with the African being "the *reverse* of the European." Weak-willed, easily reconciled to servitude, careless, exceedingly simple, "born for submission," essentially "*grands enfans*" (large children), Africans betrayed, to the scientific eye, "a degradation towards the *ape genus*, and should their appearance not betray such a degradation, their moral character would show it sensibly." In addition to other telltale signs of inferiority, he noted, Africans, "on account of the projection and inclination of their teeth, cannot pronounce the letter R."[22]

Take it all in all, Virey asserted, "it is . . . impossible to deny that [Africans] form not only a *race*, but truly a *species*, distinct from all other races of men known on the globe."[23] To clinch the argument, he included an illustration (that did not appear in the English translation) that clarified the linear progression of Camper's facial angle from simian to Negro to the Apollo Belvedere—who, to underscore the point, is actually given a slight overbite. Nott was so impressed by Virey that he sought out the French edition and reproduced a version of this illustration, with the overbite corrected, in what would become his magnum opus, *Types of Mankind*.[24]

While there was much for Nott to admire in both Long and Virey, however, there was also cause for frustration or impatience. Both writers had argued for a species distinction between Caucasians and Negroes, but neither had accounted for the origin of those differences in a way that satisfied Nott. Long had alluded to the possibility that the Deity "might, if it was his pleasure, diversify his works," while Virey had suggested that all species in the human genus had a common origin, with differences appearing long after the moment of creation. For Nott, the first suggestion was insulting to God, who was represented as an idle doodler, and the second was insulting to Caucasians, who were crowded into kinship with their natural inferiors. And the idea that the corruption of the white

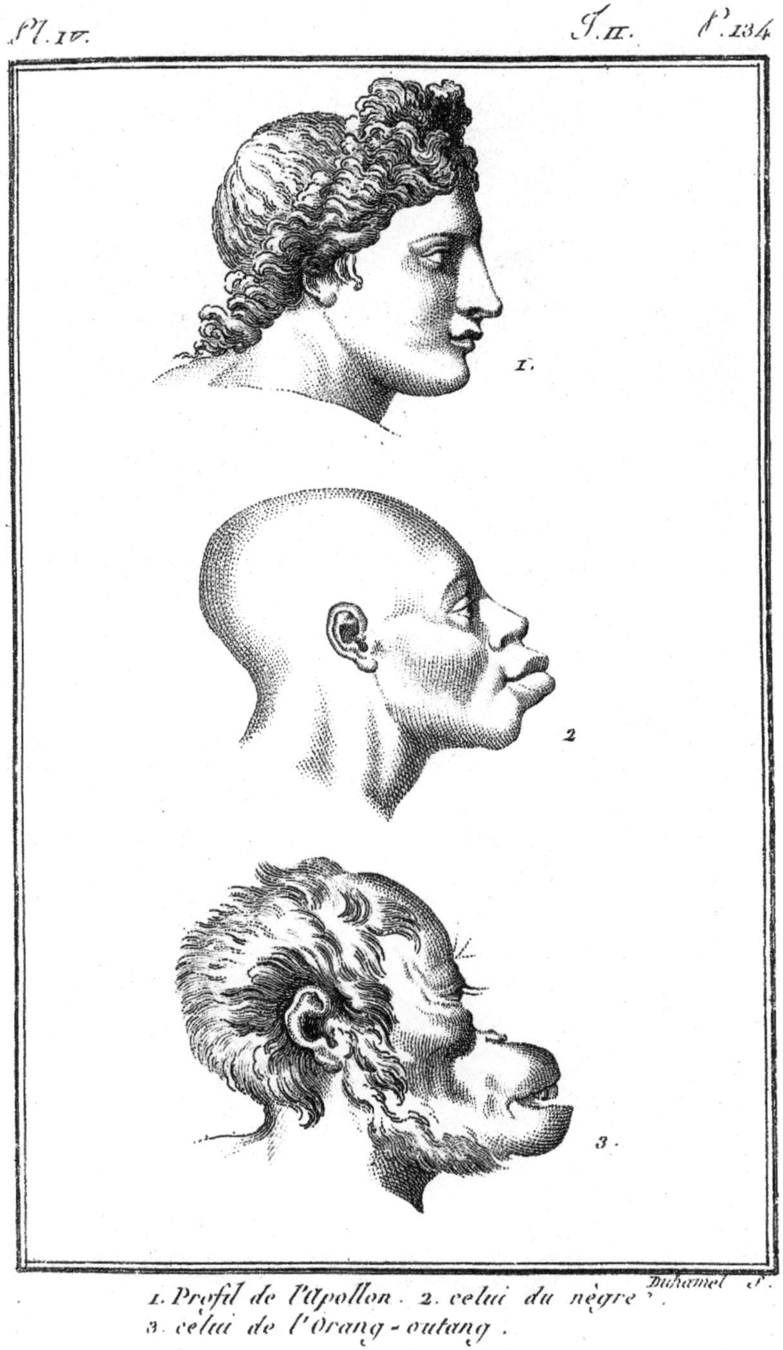

FIGURE 3.5. Duhamel, "The Facial Angle of Species," in *Histoire naturelle du genre humaine*, by Julien-Joseph Virey (1801), 2. Not reproduced in English edition (1837) translated by J. H. Guenebault. Public domain. Source: Wikimedia Commons.

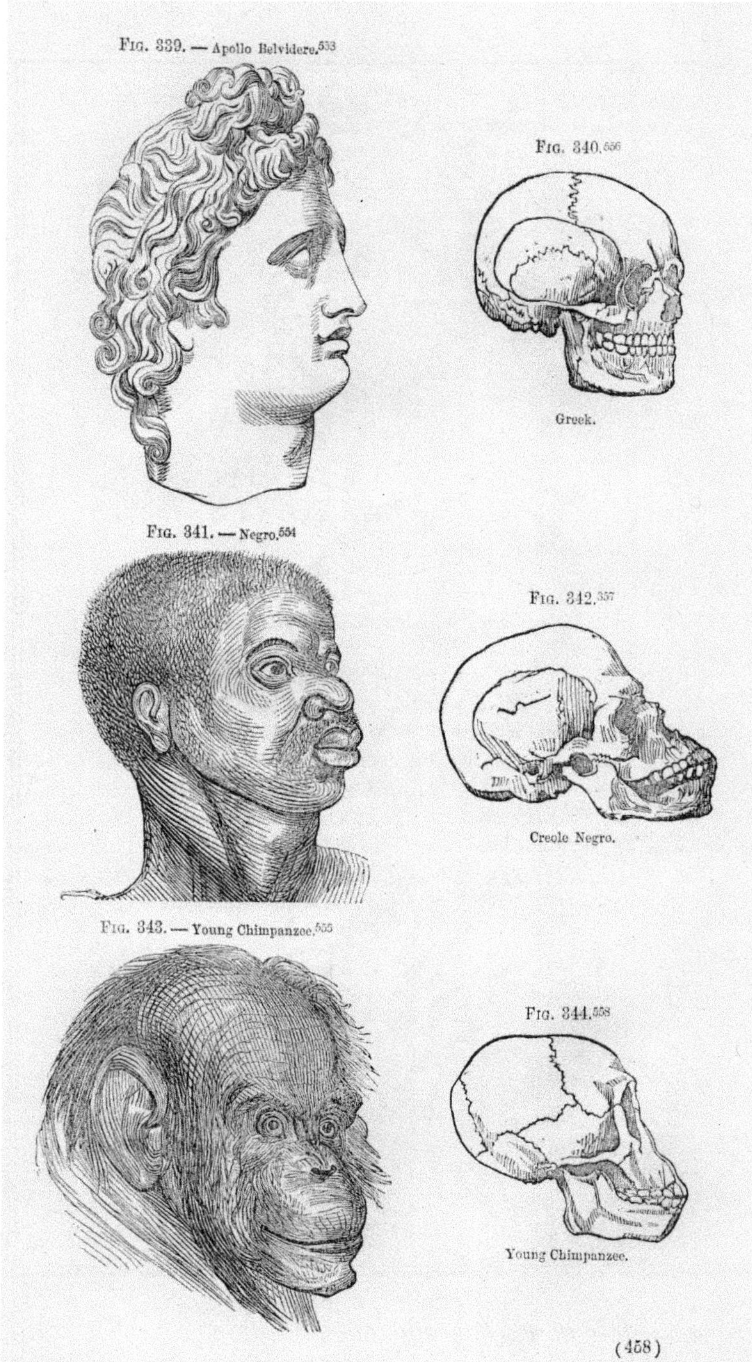

FIGURE 3.6. From J. C. Nott and Geo. R. Gliddon, *Types of Mankind* (1854), 458.

race—really, of the concept of race itself—by miscegenation could simply be washed away in a few generations was of course unsound.

Nott had something far bolder in mind. The "natural diversity of the human intellect" that Long posited was, Nott declared, no accident. The novel truth—which Nott would be the first in America to proclaim—was that God created the various human species, along with other fauna and flora, whole and complete, disposing them around the world both before and after the Deluge, in environments suited to their endowments.[25]

This elegant solution, supported both by Christian faith properly understood and scientific observation, did not, however, sweep away all opposition. In the 1840s, few of those who believed in God believed that God had done what Nott said he had. The Genesis account had long been accepted as a lucid and deeply satisfying explanation of human origins. There was little reason or incentive, many felt, to question the ancient idea of a primal pair, the ancestors of all humankind, or to look beyond the sons of Noah for the origins of human diversity. In order to win over adherents to traditional beliefs, Nott had to confront and defeat the established reading of the Bible as a literal description. This was not a question to be decided by pointing to the plain fact of difference. This was a question of how to understand the world, beginning with a correct understanding of the Word.

A committed if idiosyncratic Christian affiliated with no denomination, Nott was an equally ardent advocate for modern science. He saw no conflict between these commitments, and in fact believed that science provided superior explanations of events rendered in the Bible in a mythical or fabulistic manner. Those who preferred the murky comforts of a traditional understanding of Scripture to the cold, clear, invigorating sunshine of the truth were, he held, simply not advantaging themselves of the benefits of modernity. True enlightenment would begin when the Bible was understood not as the Word of God but as a corrupted text recording the origin story of a single group in figurative language that had passed through the hands of transcribers, translators, and editors for over two thousand years. The universe was created by God but the Bible was written by man, and it was scientists who had to assume the responsibility for saying so.

Few scientists chose to accept this responsibility. Among those who took up the subject of racial difference, neither Camper, Herder, Samuel Stanhope Smith, Blumenbach, nor Morton had been willing or able to establish a separate domain of science where the truth of the Bible did not pertain. Even Sir William Jones, one of the most radically original thinkers of the late eighteenth century, had been limited in the conclusions he had been able to draw by the premise that the "Mosaic text" was "more than human in its origins." Having extended the reach of history dramatically by identifying a common origin of Latin, Greek, and Sanskrit in a now-lost tongue or family of tongues that would come to be called "Proto-Indo-European," Jones had then undercut his own scholarship and scholarship in general by identifying the three primary linguistic groups as descendants of the sons of Noah, "the only human family after the flood."[26]

When treated in this way as the horizon of knowledge, the Bible, Nott believed, actually blocked the inevitable progress of man as a creature capable of understanding and self-understanding. To think, as Jones and nearly all other Christians did, that humankind was a single immense family was to remain in an intellectual prison of one's own making, favoring inherited dogma over science and plain observation. The Bible must be read differently, the world must be understood differently, the human condition must be seen differently, and not until all this happened could the epoch truly be called enlightened. The times called for bold statements.

"The Mulatto a Hybrid" had been published in 1843. Buoyed by the recognition it gave him, Nott became (even) less circumspect. In the winter of 1843–44, he gave two talks to the Franklin Society in Mobile that made the earlier essay seem a model of argumentative decorum and probity. The transcripts, published as *Two Lectures on the Natural History of the Caucasian and Negro Races*, made his reputation as a scientist, polemicist, and fearless theological contrarian. Comparing himself to Galileo, Nott stated his revolutionary case with defiant clarity: "There is a Genus, Man, comprising two or more species"—a question, he assured his readers, of "deep Political, Moral and Religious import."[27]

The Egyptian Origins of the American School

Nott's lectures, along with his essay on the mulatto, met with predictable outrage from Northern abolitionists, who regarded the whole parade of argumentation he had marshaled as a screen for and distraction from the proslavery, home-rule, state-sovereignty agenda of the militant South. Those and others were offended by Nott's pugnacious disrespect for the traditional reading of the Bible and his disregard of such bedrock articles of faith as the sense of a shared human ancestry, the common human burden of original sin, and the universal significance of Jesus as redeemer. He was losing the argument even among the proslavery advocates in the South.

In the scientific community even in the North, however—even, eventually, in Europe—Nott was taken as a serious man whose conclusions commanded respect not only because they were supported by conventional wisdom about Africans but also because they reflected advanced thinking, which was increasingly drawn to the concept of "diversity" in the origins and aptitudes of different races. Nott's public reflections on diversity, which he referred to in private letters as his "niggerology," brought him to the attention of the world. In 1866, shortly after the defeat of the Confederacy, an English writer described Nott—who had by then dined with Morton, Louis Agassiz, and Charles Lyell—as "the greatest living anthropologist in America."[28]

Simple enough in their essence to be readily understood by non-scholars, Nott's lectures included observations that would have drawn a knowing smile from a popular audience, such as his reference to the "*bouquet odour* of a negro's skin."[29] But they also had sufficient complexity and detail to impress more sophisticated intellectual tastes. In a compact but wide-ranging discourse that embraced anatomy, biology, theology, anthropology, natural history, and ethics, Nott provided controversial but compelling answers to many of life's persistent questions.

Doubts about the earth's antiquity had been sown by discoveries of vast geological periods that could not be accommodated to the 6,000-year chronology of creation established by Archbishop James Ussher in 1650.[30] While many people were beginning to accustom themselves to the idea that the universe was far older than Ussher had claimed, a significant number even among the educated classes still held to the belief that humankind was of what now seems quite recent origin, a consensus that did not crack until the publication of Darwin's *On the Origin of Species* in 1859 and Lyell's *Antiquity of Man* in 1863.[31] Committed to the scientific perspective and impressed by recent research that showed that a pyramid-building culture existed in Egypt less than a century after Ussher's date for the Deluge, Nott was predisposed to join in the critique of Ussher's chronology. If the authority of the Bible on this point could be questioned, then Bible-based monogenesis would be weakened even as the moral authority of revealed religion would be retained. Nott would be seen as a rationalist, a scientific Christian, and people would feel they were being persuaded by a distinguished intellect representing a modern form of traditional faith rather than being seduced by a rank atheist. The path would be cleared for science, plural creations, species differences between blacks and whites, and by implication the slave culture of the South.

The immediate source for the information about Egyptian art and society that Nott cited in his essay on insurance rates was Morton's *Crania*

Americana. But Morton himself was drawing on the work of George R. Gliddon, a flamboyant personality who had spent much of his life in Egypt, serving ultimately as American consul in Cairo, where he had taken an intense interest in Egyptian hieroglyphs, obelisks, tombs, monuments, inscriptions, mummies, and especially skulls. After decamping for the United States in 1840, Gliddon supported himself by writing about Egypt and giving public lectures featuring a spectacular "Panorama of the Nile," a long painted scroll that could be unwound on rollers and that depicted paintings, jewelry, hieroglyphs, papyri, maps, and plates illustrating the art of embalming; for a thrilling conclusion, he displayed four mummies.[32] William Stanton provides a memorable description of Gliddon as a "name-dropper, a sponger, a swinger on the shirttails of the great, a braggart, pretender, and scatologist. . . . With delusions of profundity, his mind was as shallow as a mountain stream, and, except when clouded by his turgid rhetoric, as clear."[33]

In his public presentations—the first in the United States on ancient Egypt—Gliddon focused on his startling discovery that two thousand years before Christ, when Egyptians were writing in hieroglyphics, discovering the arch, constructing massive tombs and porticos, and building monuments that still excite the world's admiration, their country was populated by two distinct races, which he could identify because they looked then just as they do now: the people creating civilization were recognizably Caucasian, while their slaves and servants were invariably Negroes, the former being, as he wrote, "as distinct from the Egyptian natives, as an Anglo-Saxon is from a Chimpansee."[34] Even before coming to the United States, the enterprising Gliddon had been corresponding with the information-hungry Morton, managing to impress the distinguished scientist by his research, persuade him by his conclusions, and place him in his debt by providing ninety-three Egyptian crania, a fabulous trove that gave Morton's study, like Blumenbach's, a Golgotha aspect.

Everything about this relationship was useful to Nott. From Gliddon Nott got the antiquity and durability of races, the Caucasian and Negro races in particular; from Morton he got scientific and empirical confirmation of the superiority of the former to the latter. These gifts arrived in the same bundle. The 1839 publication of Morton's *Crania Americana* immediately established Morton's reputation in the United States and Europe as the most rigorous and original practitioner of craniology in the world. And, because Morton so profusely thanked Gliddon in the preface for his zeal in procuring skulls, it had the same effect on Gliddon's reputation, taking him out of the clamorous vulgarities of popular entertainment and

FIGURE 3.7. George R. Gliddon, 1853. Source: National Museums Scotland, Edinburgh.

insinuating him into the quiet rooms where the real work of science was being pursued.

Crania Americana was in every way a revolutionary event, transforming a practice of skull measurement with limited means, few practitioners, and modest aspirations into something far more consequential. What Morton claimed, in more compelling detail than any previous practitioner, was that careful measurements conjoined with ethnological descriptions of peoples could become the source of literally hard evidence not just about cultural differences but about the reasons for these differences. Using sophisticated instruments of his own devising prominently including the "facial goniometer," Morton undertook a deeply original study of shapes and angles—not only Camper's facial angle, but the longitudinal, parietal, vertical, and zygomatic diameters, the occipito-frontal arch, the inter-mastoid arch and line, the sub-coronal region, and the horizontal periphery. To these he added a different kind of statistic, the internal

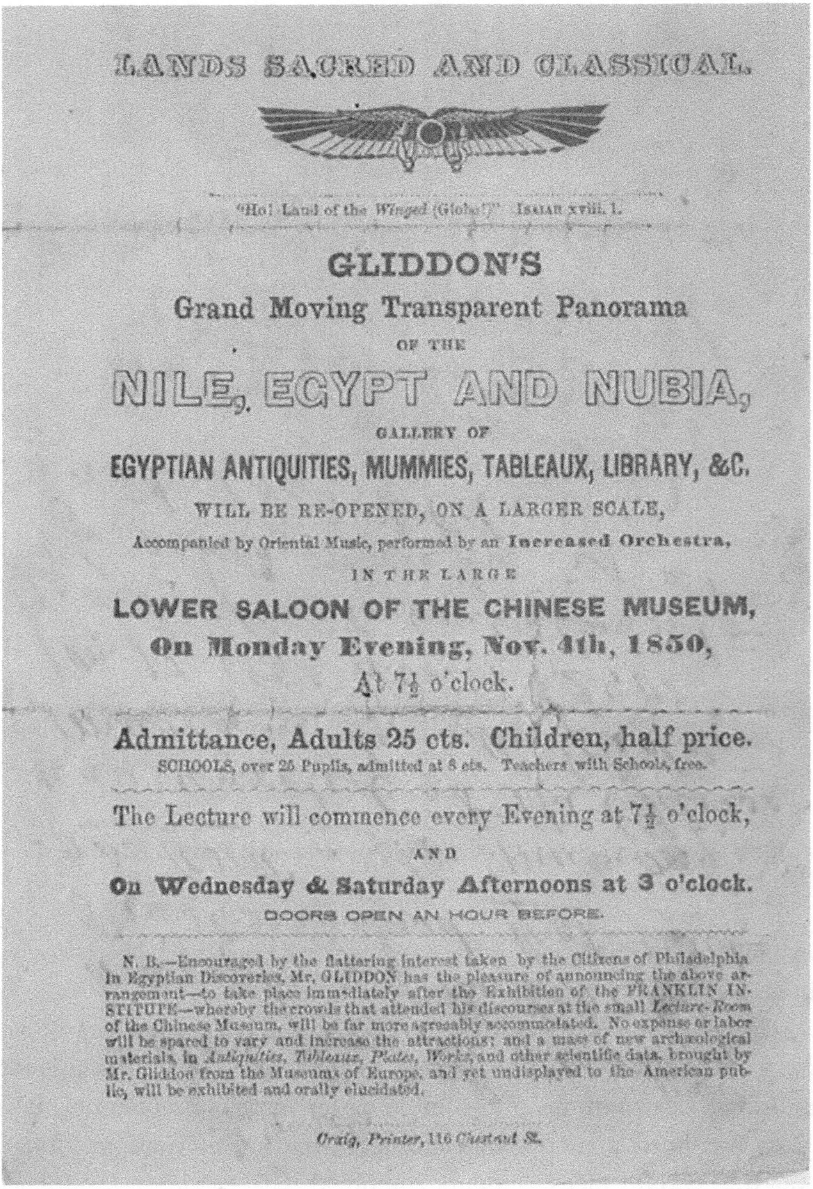

LANDS SACRED AND CLASSICAL.

"Ho! Land of the *Winged* Globe!" ISAIAH xviii. 1.

GLIDDON'S

Grand Moving Transparent Panorama

OF THE

NILE, EGYPT AND NUBIA,

GALLERY OF

EGYPTIAN ANTIQUITIES, MUMMIES, TABLEAUX, LIBRARY, &C.

WILL BE RE-OPENED, ON A LARGER SCALE,

Accompanied by Oriental Music, performed by an **Increased Orchestra**,

IN THE LARGE

LOWER SALOON OF THE CHINESE MUSEUM,

On Monday Evening, Nov. 4th, 1850,

At 7½ o'clock.

Admittance, Adults 25 cts. Children, half price.

SCHOOLS, over 25 Pupils, admitted at 8 cts. Teachers with Schools, free.

The Lecture will commence every Evening at 7½ o'clock,

AND

On Wednesday & Saturday Afternoons at 3 o'clock.

DOORS OPEN AN HOUR BEFORE.

N. B.—Encouraged by the flattering interest taken by the Citizens of Philadelphia in Egyptian Discoveries, Mr. GLIDDON has the pleasure of announcing the above arrangement—to take place immediately after the Exhibition of the FRANKLIN INSTITUTE—whereby the crowds that attended his discourses at the small *Lecture-Room* of the Chinese Museum, will be far more agreeably accommodated. No expense or labor will be spared to vary and increase the attractions; and a mass of new archæological materials, in *Antiquities, Tableaux, Plates, Works*, and other scientific data, brought by Mr. Gliddon from the Museums of Europe, and yet undisplayed to the American public, will be exhibited and orally elucidated.

Craig, Printer, 116 *Chestnut St.*

FIGURE 3.8. Handbill for George R. Gliddon's "Panorama of the Nile." Public domain.

capacity of the skull, which he determined by pouring seeds, and then, in a quest for greater precision, lead shot into them.[35] The results were recorded in *Crania Americana*, where they became part of the public record, partial and preliminary but significant contributions to the factual base of knowledge about the human species.

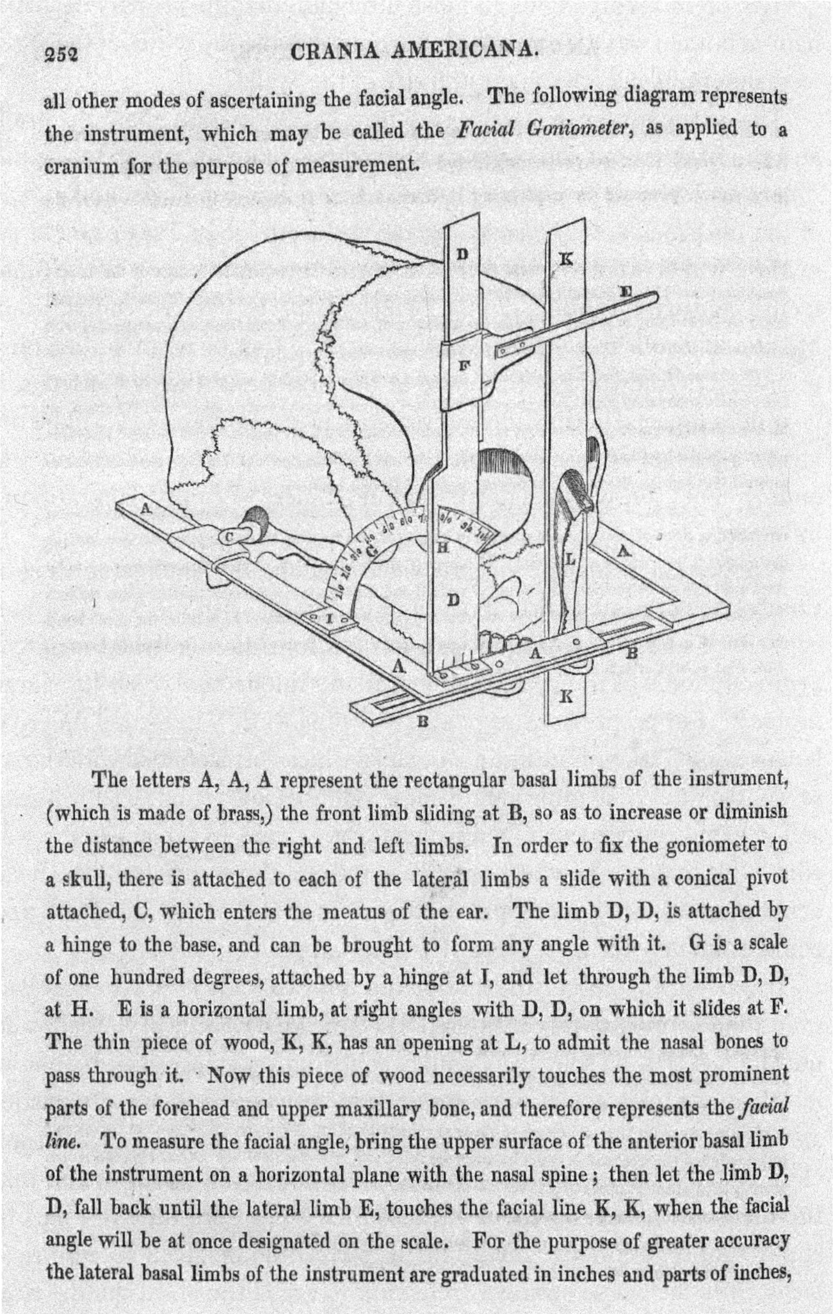

252 CRANIA AMERICANA.

all other modes of ascertaining the facial angle. The following diagram represents the instrument, which may be called the *Facial Goniometer*, as applied to a cranium for the purpose of measurement.

The letters A, A, A represent the rectangular basal limbs of the instrument, (which is made of brass,) the front limb sliding at B, so as to increase or diminish the distance between the right and left limbs. In order to fix the goniometer to a skull, there is attached to each of the lateral limbs a slide with a conical pivot attached, C, which enters the meatus of the ear. The limb D, D, is attached by a hinge to the base, and can be brought to form any angle with it. G is a scale of one hundred degrees, attached by a hinge at I, and let through the limb D, D, at H. E is a horizontal limb, at right angles with D, D, on which it slides at F. The thin piece of wood, K, K, has an opening at L, to admit the nasal bones to pass through it. Now this piece of wood necessarily touches the most prominent parts of the forehead and upper maxillary bone, and therefore represents the *facial line*. To measure the facial angle, bring the upper surface of the anterior basal limb of the instrument on a horizontal plane with the nasal spine; then let the limb D, D, fall back until the lateral limb E, touches the facial line K, K, when the facial angle will be at once designated on the scale. For the purpose of greater accuracy the lateral basal limbs of the instrument are graduated in inches and parts of inches,

FIGURE 3.9. The "facial goniometer," from Samuel George Morton, *Crania Americana* (1839).

Two premises drove this ghoulish drudgery: that the primary determinant of human variation was the brain, and that the capacities of the brain were determined by the shape and size of the skull.[36]

The argument for the skull as a source of information about the brain and the brain about the race was spelled out in an appendix to *Crania Americana* written by George Combe, a Scottish phrenologist and author of the 1824 *Elements of Phrenology* (as well as the 1847 *The Constitution of Man*, which eventually became the third best-selling book of the nineteenth century, outselling *On the Origin of Species*), whose contribution Morton had solicited. "If this view be correct," Combe wrote of Morton's premises, "a knowledge of the size of the brain, and the proportions of its different parts, in the different varieties of the human race, will be the key to a correct appreciation of the differences in their natural mental endowments, on which external circumstances act only as modifying influences."[37]

Size was the primary indicator, since small brains could, according to Combe, be correlated with idiocy, and big brains with mental vigor and capacity. As he put it, "Individuals and nations distinguished for great aggregate force of mind, animal, moral and intellectual, have had large brains."[38] But proportions were also revealing as they indicated different balances and ratios of the brain's "organs," which corresponded with thirty-seven "faculties" including amativeness, adhesiveness, sublimity, firmness, self-esteem, cautiousness, benevolence, hope, and so forth. Combe was convinced that an analytical practice of "reading" the surface of the head could be applied to nations and racial groups as well as to individuals, and could ultimately form the basis of a "diagrammatics of race."[39]

It was to Morton's credit that he did not explicitly endorse what was even then coming to be considered a soft-headed approach to the brain, but the fact that he solicited Combe's contribution at all suggests that he felt the need to establish some connection between his mass of statistics and group characteristics of the kinds reported by ethnologists and travelers. Morton believed that such a connection could be drawn, and that the hierarchical arrangement of races that was so prominent a part of contemporary ethnology including his own could be given an empirical basis. Still, he was so strangely reluctant to argue the point that he confined the most telling evidence for this connection to a suggestion in a footnote—perhaps the most consequential footnote in the history of that scholarly artifact—where it was recorded that on the basis of the skulls "in my possession, and such also as I could obtain from my friends" (Gliddon), Caucasians had the largest brains, followed by Mongolians, Malays,

RACES.	No. of skulls.	Mean internal capacity in cubic inches.	Largest in the series.	Smallest in the series.
Caucasian.	52	87.	109.	75.
Mongolian.	10	83.	93.	69.
Malay.	18	81.	89.	64.
American.	147	80.	100.	60.
Ethiopian.	29	78.	94.	65.

FIGURE 3.10. From Samuel George Morton, *Crania Americana* (1839), 260, footnote.

Americans, and Ethiopians.[40] He did not, in this note, explicitly connect brain size to intelligence or general aptitude but for many readers the connection went without saying.

The most immediately striking feature of *Crania Americana* was undoubtedly a series of seventy-one haunting and evocative lithographs of skulls and mummified remains, but the argument of the book was carried by Morton's ninety-five-page introduction, "On the Varieties of the Human Species." In a series of lively and detailed accounts, Morton described the five major races and their twenty-two subordinate "families," groupings that he explicitly stated "are not assumed as identical with races." The Circassians, he noted, were "celebrated for superior personal endowments"; the Abyssinians were "liars, as well as the Arabs; but they yet have a feeling of shame which the Arabs have not"; the Bosjesmans were "far more savage and degraded than any other Hottentot tribes. . . . They are robbers by profession, cruel by nature." For Morton, such characterizations, with their inbuilt value judgments, could coexist with measurements as complementary ways of knowing, so that, for example, the unequaled "gluttony, selfishness, and ingratitude" of the Eskimaux of Prince Regent's Bay could be correlated with the prominence of their cheek bones.[41] What made all this not just picturesque but scientifically interesting was the correlation between cranial capacity and cultural traits, without which *Crania Americana* was just a travelogue with skulls.

Morton's work was received even in Europe as a scientific breakthrough, a eureka moment of sudden insight. Praised by James Cowles Prichard as a dramatic advance, *Crania Americana* represented for many the triumph of the empirical method, a compelling demonstration of how a field previously dominated by narrative, description, and speculation might yield to the cold probing eye of science.[42] But for Morton himself, it was Gliddon's insight into the culture of ancient Egypt that provided the breakthrough.

Morton was so aroused by Gliddon's discovery that he launched another footnote, this one epic in proportions, so sprawling and complex it spawned footnotes of its own. In the course of its eleven hundred words on "the Supposed Affinities between the Egyptians and Negroes," Morton cited one authority who remarked that he had seen on the monuments "many decapitated figures: these were all dark, while those who had struck off their heads, and still stood over them sword in hand, were red."[43] The color might have been an impediment to identifying the executioners as Caucasians, but Morton assured the reader that, "It is easy to prove that *whatever may have been the hue of their* skin, they belong to the same race with ourselves." They look like us, and the servants and slaves look like Negroes; "the eye," Morton said, "can mostly detect them." And if the eye detected Negro features in the Sphinx, the most likely explanation was that that famous structure was "a shrine of the Negro population," which had, presumably during periods of rest generously granted by the Pharaoh, built its own monument.[44]

Morton's discussion of "the Nilotic family" affirmed in a more professional discourse Gliddon's assertions about the racial composition of ancient Egypt. This was very much to Nott's purpose, for it provided the best possible explanation for something Nott was eager to explain, the basis in nature of the master-slave relationship that prevailed in the Southern states. On Morton's authority, Nott could claim that from the most ancient times, Caucasians had borne responsibility for the heavy lifting of governance, culture, religion, and science while Negroes had been charged with the heavy lifting of lifting.

If any were uncertain why Negroes in the South, who had been in the United States for over two hundred years, had failed to make more than modest progress in manners and morals, learn to survive in cold weather, or master more than the rudiments of self-governance, the mystery was solved in a stroke: their brains were just too small and the relations between the faculties unpropitious. And if any readers were not appropriately alarmed about the consequences of miscegenation, they could look to the sorry state of modern, mongrelized Egypt, with its thoroughly corrupted Coptic population degraded by millennia of mixing with inferior races.

Without implicit support from Morton, Nott might well have been dismissed in the North and in Europe as a mere polemicist for the slave-holding South, which he certainly was, increasingly so in the years leading up to the Civil War. But with that support, Nott—still widely distrusted and even reviled by Northern abolitionists—was able to persuade many that his arguments were legitimate, good-faith scientific conclusions with important political implications to be sure, but no political determinants.[45]

Crania Americana may have been useful to Nott, but it was still, to Nott's bolder temperament, marked and marred by timidity. Morton would, for example, note that Blumenbach compared Negro skulls in his possession to those of primates; and he would cite Camper's useful statement that the facial angle extended from the "the sublime beauty of the ancient Greek models" down to the Negro, below which "we have an orang outang, or a monkey." ("If we descend still lower," he added, "we have a dog or a bird.")[46] Morton was even willing to rank races or families within races, placing, for example, the "Negro family" at the top of the "Ethiopian Race" and the "Alforian Family" at the bottom, at "the lowest grade of humanity."[47] But he seemed unwilling to progress from learned references to conclusions about larger and more important issues. He noted the resemblances between Negroes and monkeys but for some reason could not bring himself to state what was obvious to Long, Virey, and Nott himself, that Negroes were different from Caucasians not just in appearance but in status in the order of creation.

Nott could do better. Citing statistics, he said plainly what Morton had all but said: they are more like apes than they are like us, and are sufficiently unlike us for the scientist, acting in the interest of accuracy, clarity, and respect for God's original design, to place them in their own zoological category. Everything argued for it: the conformation of the head, the projecting upper jaw, the deficiency in the area of the chin, the extraordinary length of the arm, the fullness of the muscles on the posterior of the pelvis, the obtuse angle of the bones at the knee joint—all of which constituted "points of resemblance . . . between the Negro and the Ape." To an anatomist, the difference between the Negro and Caucasian is "greater than the difference in the skeletons of the Wolf, Dog and Hyena, which are allowed to be distinct species."[48] Empirical truth directly stated.

If Morton was unwilling at this point to concede that Negroes and Caucasians were different species, he was more than prepared to describe racial diversity as a fact of nature. In an 1842 lecture at the Pennsylvania Medical College, the now-famous Morton told the students that human variation was evidence of an "all-pervading design" that had placed species in environments suited to their capacities. The "wretched and brutalized

Hottentot" had been deposited in the warmer latitudes where he could survive, while the mentally and morally superior Caucasian had been given provenance over "those happier climes which lie north of the tropic of cancer." Primordial attributes of mind were responsible for the fact that civilized Europeans had been able to repulse the Mongolians who had sought to exterminate them, just as they were now in the process of seizing "these happy climes which we [Americans] now inhabit" from the "wild and untutored Indian." In fact, Morton told the students, Caucasian history constituted a steady advance "from one degree of cultivation and refinement to another, which the other races have never approached, nor are likely, in the ordinary course of events, ever to realize."[49]

The phrase "ordinary course of events" sounded the note that Nott wanted to hear, that Caucasian superiority and Negro inferiority were part of God's plan and could only be compromised by the man-made disaster of unchecked (and extraordinary) racial mixing. When Nott read Morton's lecture, he could see the differences between himself and Morton shrinking as Morton concentrated his attention on the racial extremes of Caucasians and Negroes and adopted a judgmental vocabulary in which approval, admiration, wonderment, disgust, and shame were advanced as scientific facts. Indeed, while Morton may have been unaware of changing his positions in any significant way, he was over the course of the 1840s aligning himself by degrees, in the manner of a dull blade being slowly sharpened, with the agenda and style of the younger, less distinguished, less generally admired, more polemical, more political, more nakedly ambitious Nott.

This alignment is evident in Morton's 1844 *Crania Ægyptiaca*, a less imposing and more directly Gliddonesque volume—indeed, it was dedicated to Gliddon—than its predecessor but valuable for Nott in that it stated, more forcefully than *Crania Americana*, the primordiality of the master-slave division of the races. Working with Egyptian skulls so ancient they had the consistency of baked clay and sometimes shattered during transport, Morton catalogued, measured, and analyzed each one, giving its provenance (sarcophagi, pyramids, grottoes, or mummy pits) as well as the age, sex, and likely status of the person: "Four Heads from the Grottoes of Maadbdeh," "Four Heads from a Tumulus," "Two Heads of Lunatics, from Thebes." He was, he said, pursuing a project that Blumenbach had sketched out half a century before: "a very careful, technical examination of the skulls of mummies." But he was adding an emphasis Nott embraced but Blumenbach would have rejected by focusing on "two of the great races of men, the CAUCASIAN and the NEGRO."[50]

What had been one strand among many in *Crania Americana* became the dominant theme of *Crania Ægyptiaca*, and it was Nott's theme, describing Nott's America. Ancient Egypt, Morton wrote, exhibited perfect racial clarity in which there were "no gradations . . . between barbarism and refinement." The Negroes depicted in the paintings are "precisely as we are accustomed to see them in our daily walks: indeed, were we to judge by the drawings alone, we might suppose them to have been executed but yesterday; and yet some of these vivid delineations are nearly three thousand five hundred years old! And, moreover, as if to enforce the distinction of race by direct contrast, they are placed side by side with people of the purest Caucasian features." In this respect, Morton observed, little has changed in the thirty-five centuries since Egypt had created civilization: Caucasians rule, and while "Negroes were numerous in Egypt . . . their social position in ancient times was the same that it now is, that of servants and slaves."[51]

Morton's at least implicit endorsement was indispensable for Nott; and yet from Nott's perspective, the partially disabled Morton—his health was always uncertain, and a collapsed lung had left him with a lowered shoulder and limping gait—still lacked the required backbone. Not only had Morton gotten on his high horse to denounce the "degradation" of slavery, but he had committed scientific sins, at least of omission. He had measured the internal capacity of skulls, but had left it to Combe to put the sharpest point on the underlying premise of such work, that brain size correlated with intelligence and other virtues.[52] Moreover, Morton tended to swaddle his assertions in qualifications, adopting a rhetoric of speculation where direct statements were clearly justified. Rehearsing for the benefit of the medical students in his 1842 lecture the triumphant spread of Anglo-Saxon dominance across the globe, Morton said that he had become convinced that the "primeval differences" among men "have not resulted from physical causes acting on constitutions originally the same," and were "not an accidental occurrence"—but said no more.[53]

Not this, not that, but no positive statement about God's will or the fundamental rightness of the entire plan, nothing more than a platitude about the vague "necessity" governing the "economy of creation." To Nott, Morton seemed more interested in questions than solutions, more inclined to proliferate doubts than to produce conviction.

Even worse, at certain points in *Crania Americana* Morton would actually undercut himself, speaking against the very point he seemed to be making. The skulls of American Indians, he noted, had a mean internal capacity of just eighty cubic inches, barely outpacing Ethiopians, and their intellectual capacity appeared to be "of a decidedly inferior cast," and yet "every one knows . . . that the mind expands by culture; nor can we yet tell how near the Indian would approach the Caucasian after education had been bestowed on a single family through several successive generations."[54]

Surely, Nott may have thought, maybe not "every one," but certainly *every reader of Morton* would know that races are fixed in their capacities, and that a few generations of education would not transform a race that had for thousands of years been servants, slaves, and bearers of tribute into self-governing citizens of a modern society. And when the softly sympathetic Morton wrote that "the advantages of education have been inadequately bestowed" on Negroes so we did not actually know what level they might attain under different circumstances, Nott may have wondered why Morton was measuring skulls at all if they yielded no information about the brain's capacity to be educated.[55] (And if they did yield hard information about a fixed capacity, as Morton had seemed to believe, then why waste the bestowing?)

From Nott's perspective, Morton was entirely too fond of conventional but unscientific phrases such as "the human species," or even worse, "the human family." But perhaps the most misleading and even possibly heretical single opinion in *Crania Americana* from Nott's point of view was Morton's deeply unfortunate comment that "slavery has degraded the faculties" of both present Mexicans and Copts to the point where "it would require centuries of the most favorable circumstances to resuscitate the dormant genius of either."[56] To Nott, such hopeful pieties violated the principle, which Morton elsewhere seemed to support, of the permanence of racial characteristics and reflected a muddleheaded philanthropy that could content itself with vague allusions to a dreamy egalitarian future in which history was obliterated in a general awakening of long-dormant genius.

The problem was that like so many others, Morton was clearly under the spell of the traditional reading of the creation story in the Hebrew Bible, which Morton had invoked in a singularly flaccid passage in the introductory essay in *Crania Americana* where he took up the question of "the origin of the human family." The prevalent belief, Morton had noted in his 1842 address at the Pennsylvania Medical College,

> is derived from the sacred writings, which, in their literal and obvious interpretation, teach us that all men have originated from a single pair; whence it has been hastily and unnecessarily inferred, that

> the differences now observable in mankind are owing solely to vicissitudes of climate, locality, habits of life, and various collateral circumstances. . . . [W]e may inquire, whether it is not more consistent with the known government of the universe to suppose, that the same Omnipotence that created man, would adapt him at once to the physical, as well as to the moral circumstances in which he was to dwell upon the earth?[57]

We may inquire? To Nott, the inquiring season had passed and the answer was plain: Omnipotence had created human beings adapted to specific environments, and Negroes, helpless in the cold, flourished in heat. Morton knew this and yet for some reason had felt compelled to muddy the waters by noting the irrelevant fact that "real Negroes constitute the indigenous population of Van Diemen's Land, which is nearly as cold as Ireland."[58] With a disturbing frequency, Morton seemed uncertain whether he was speaking to fellow scientists or fellow communicants in the traditional faith, as when he mentioned "the descent of mankind from the ark" or identified ancient Egypt as "the intellectual centre of the posterity of Ham."[59]

It has been customary to think of the older and more distinguished Morton as Nott's intellectual as well as social superior, but in fact the influence flowed in the opposite direction. Throughout the collegial personal relationship that developed between the two men during the 1840s, it was Nott whose positions were fixed and specific, and Morton who seemed to be drawn almost magnetically in the direction of Nott both in the subjects he took up and the arguments he was making.[60] While Nott did not presume, in their friendly, mutually reinforcing communications, to contradict Morton directly, he did manage to bring Morton along through a combination of flattery, encouragement, repetition, and the repeated assurance that they were at bottom kindred spirits, seekers after truth. Above all, Nott wanted Morton to agree, or at least not to disagree, with the critical propositions that the Bible was an unreliable guide to history, that the races had been created separately and placed in the parts of the earth where they were equipped to flourish, that science could specify the differences in capacities among the races, and most important, that racial distinctions between Caucasians and Negroes were species distinctions.

Morton was plainly inclined to agree with all these positions, but his views never quite assumed the crystalline forms Nott preferred. Morton's deliverances exploited the capacity of language to express nuances and shades of meaning. He spoke, for example, of an "original diversity of mankind" or "*primeval diversities* among men"; he even described races as "primordial organic forms."[61] But these phrases allowed for a single complex creative act, which would preserve a sense of a common origin and weaken the case for species differences. Nott wanted a plain statement of the deliberate creation of separate and distinct species.

On January 29, 1850, Morton finally rewarded Nott's forbearing patience.

After reading *Two Lectures*, Morton wrote to congratulate Nott on the triumphant manner in which he had refuted the case that one race might be transmuted into another.[62] Defending but also confessing to his own hesitation in making this case so boldly, he explained that he had feared the consequences of a "controversy with the clergy." In fact, one of those clergy, Dr. John Bachman, a slave-owning Lutheran pastor in Charleston, had been for some time peppering Morton with letters and published statements in which he asserted the authority of the Bible, monogenesis, the unity of mankind, the fertility of mulattoes, the conspicuous lack of Mortonian "repugnance" among Caucasian men for Negro women, the improbability of a newly created aboriginal race inhabiting the Arctic, and the overall superiority of white people. Bachman had framed his statements in such belligerent and condescending terms that the equable Morton, who had been reluctant to engage in any unpleasant public disputes, had at last been provoked to respond with a direct statement of the innate and unalterable differences among the races. Somewhat to his surprise, Morton had found the experience fortifying, even invigorating.[63] And so, he told Nott, he was at last prepared to embrace the truth as Nott had stated it.

And there it was, near the end of the letter, the concession Nott had been seeking for so long:

> ... how beautiful, how evidently and inherently truthful is the proposition—that our species had its origin, not in one, but in several or in many creations.... Here is the truth divested of mystery; a system that explains the otherwise unintelligible phenomena so remarkably stamped on the races of men.[64]

Morton had arrived at the bourne of polygenesis—a clear statement not merely that diversity was original but that the races themselves were the products of separate acts of creation.

A mere half-step would bring Morton to the scientifically and politically precious assertion that the two polar races of Caucasian and Negro were not only the issue of separate acts of creation but—the scientific consequence—distinct species in a biological sense. He seemed inclined to take it. In fact, shortly after writing this letter, he submitted an essay to the *Charleston Medical Journal and Review* with the title "Notes on Hybridity" in which he rehearsed the criteria for species identification, turning at the end to the question that Nott had addressed seven years earlier in "The Mulatto a Hybrid," the implications for human beings.[65] Promising a fuller exposition soon, he cited with approval W. F. Van Amringe's statement in *The Natural History of Man* that the argument for the unity of humankind based on the fact that the different races run together by gradations "amounts to nothing; for the anatomical and physiological structure and functions of the different races of men, are sufficient to constitute distinct species."[66]

Three months later, Morton suddenly died without ever having made that fuller exposition in his own voice. But he had said enough so that he could be counted a solid ally for Nott, whose fame was by that time rivaling Morton's. If the kindly, humane Morton had been troubled by the connection between Nott's scientific labors and the practice of slavery, or by the possibility that he himself might be seen as making slavery-adjacent arguments, he did not say so, perhaps unwilling to create any unpleasantness or provoke any controversies.

Nor had he raised such issues with Nott, even when Nott wrote to him boasting that his "niggerology . . . has made me a greater man than I ever expected to be—I am the big gun of the profession here."[67] Morton apparently permitted points of agreement on scientific questions to overrule or neutralize any differences they might have had on peripheral issues of tone or manner. A certain academic naïveté may have clouded his judgment, or he may have been sublimely oblivious to the way in which his association with Nott might be viewed. Shortly after his death, the *Charleston Medical Journal and Review* published a memoir of Morton that left no room for doubt on this score, with the author stating that "we can only say that we of the South should consider him as our benefactor, for aiding most materially in giving to the negro his true position as an inferior race."[68]

On Morton's death, Nott made of him a kind of trophy. With Gliddon, who was by then living in Nott's orbit in Mobile, serving as a glissando between the two men, Nott undertook the preparation of a massive composite text. Appearing the following year, *Types of Mankind* was a 738-page tome complete with maps, tables, chronologies, and hundreds of illustrations, some of them foldouts. The entirety was dedicated to Morton, and featured both a lithograph of Morton before the title page and a lengthy "Memoir" of Morton by Henry S. Patterson, a Philadelphia colleague. Drawing on many letters and personal exchanges, Patterson rehearsed Morton's intellectual career and painted a picture of a man entirely worthy of the public adulation he had received, a gentleman whose "mental constitution" was "free from all violence of passion . . . habitually cheerful, hopeful, and kindly."[69] Every effort was made to present the entirety as a fitting monument to this human exemplar.

The text was a mixed bag, almost a kind of tent show that included Patterson's forty-page memoir, a nearly four-hundred-page section primarily by Nott on racial types (but including a "monumental history of dogs"), a chapter by the Mobile physician William Usher on geology and paleontology, an essay by the young and recently-arrived Swiss naturalist Louis Agassiz, thirty pages of Morton's "inedited manuscripts," and finishing with two hundred and fifty pages by Gliddon including a ninety-page deliverance widely considered unreadable on the tenth chapter of the Book of Genesis, a discourse on Biblical ethnography, and a learned supplement canvasing archaeology, the development of writing, and human chronology.

A chapter by Nott on "Negro Types" was central to the boldly original argument of the entirety. Drawing on Gliddon's discoveries, it argued that the ancient Egyptians were the "*autocthones* of the Lower Nile," an ancient type of mankind from which other races had developed each in its own way, with the Africans descending and the "West-Asiatic" types ascending. The race commonly known as Caucasian was simply the highest form of this Egyptian type, but even in ancient times, the Egyptians were sufficiently differentiated from their African inferiors that they could hold them as slaves. Depictions of slaves were adduced as evidence for this proposition, with supporting commentary from ancient sources and modern scholarship that proved that "in the second century after c., the physical characteristics of a 'field' or agricultural, 'Nigger' were understood at Rome 1800 years ago, as thoroughly as by cotton-planters in the State of Alabama, still flourishing in A.D. 1853." The fact that slaves whose forebears were imported to the United States are acknowledged to be far more impressive specimens than their African counterparts can be attributed to "ceaseless contact with the whites"; but two generations of such contact would, according to Nott, do all that could be done to improve them. Certain forms of proximity—delicately alluded to as "ultra-ecclesiastical amalgamations"—have only "deteriorated the white element in direct proportion that they are said to have *improved* the black."[70]

All this and more was marshaled in support of the argument that races were not mere varieties but fixed and unalterable types, a term Nott defined in a way that aligned it with the concept of species, a "primitive or original form" whose distinctive features could not have been formed by climate or any external agency.[71] A truculent introduction by Nott anticipated and even welcomed criticism, almost defying the reader to challenge the volume's immense mass of erudition and argumentation in the fields of natural history, craniology, anthropology, history, geography, anatomy, ethnography, and Biblical criticism. All in all, *Types of Mankind* was a sensational, expensive, lavishly illustrated publication intended to communicate authority as the triumphant polygenetic manifesto of what Nott baptized "the American School of Ethnolology," generously, strategically, and falsely crediting Morton rather than himself as its founder.[72]

The all but explicit ambition of the American School, in which it largely succeeded, was to center the science of race in the proslavery context of the antebellum American South and to identify the issue of the species status of the Negro race as the central question.

While Nott insisted that Morton was the founder of the American School, readers of *Types of Mankind* would have encountered a number of arguments and emphases that were Nott's alone. The most dramatic of these involved not crania or studies of hybridity but theology-based metahistorical grand theorizing and prognostication. As a man of the library who had spent many long hours pouring seeds into skulls, Morton rarely if ever ventured statements of this kind or on this scale, but Nott (student of mosquitoes and leeches and advisor to insurance companies) was committed to the larger view.

Nott would pronounce, for example, with far greater confidence than Morton ever possessed on the "plan" or "design" of the Creator in distributing the various races in their own "distinct zoological provinces," where they should stay if they wished to be in conformity with God's will. He would also identify without hesitation the historical mission of the races, saying that "nations and races, like individuals, have each an especial destiny: some are born to rule, and others to be ruled. . . . No two distinctly-marked races can dwell together on equal terms. Some races, moreover, appear destined to live and prosper for a time, until the destroying race comes, which is to exterminate and supplant them."[73] While Morton had said that the Caucasian race was "distinguished for the facility with which it attains the highest intellectual endowments," he never ventured anything like Nott's statement in *Types of Mankind* that Caucasians were "destined eventually to conquer and hold every foot of the globe. . . . it is written in man's nature by the hand of his Creator."[74] Nor had Morton said, as Nott did here, that if distinctions between races were erased so that all peoples were hybrids, the result would be the extermination of the entire human family.[75] And while Morton had asserted that the "Ethiopian Race" included, at the bottom end of its considerable range, "the lowest grade of humanity," he had not drawn what was to Nott the obvious conclusion, that in the United States slavery was the use for which they were best suited.[76]

These formulations reflect both Nott's deepest commitments and his starkest differences with Morton. Optimistic about the prospects for the realization of God's zoological plan despite the confusion introduced in the species by the impious lust of fallible humans, Nott interpreted the feebleness of mulattoes and the reported mortality rate of Negroes in Boston—minimal and questionable data that he never attempted to confirm—as evidence that God's will was being realized not through Deluges or miraculous interventions but through the laws of nature operating over time. Given that the full realization of God's plan was nothing less than the perfection of the species, half-measures were not to be tolerated.

> Nature marches steadily towards perfection; . . . it attains this end through the consecutive destruction of living beings. . . . [T]he inferior races of mankind were created before the superior types, who now appear destined to supplant their predecessors. . . . [W]hatever may have been the order of creation, the unintellectual races seem doomed to eventual disappearance in all those climates where the higher groups of fair-skinned families can permanently exist.[77]

Nott's contempt for the lower orders did not reflect merely or exclusively local urgencies, or a parochial Alabama prejudice. A student of history, he attributed political unrest in France to the misguided tolerance exhibited toward the southern Gauls, who were alike unfitted for reason and for a republic. Unlike Morton, Nott did not mince words: "If the imperial government," he wrote, "were simply to chop off the head of every demagogue who was not a blond *white*-man, they might 'get along' in France as in England, Germany, and the United States."[78]

In such passages the true lineaments of the American School as Nott shaped it emerged like a photographic image in the process of chemical development. The image is now clear—but what is it? For most post–Civil War readers, the American School is a political and ideological project appearing in the costume of science. An authority on white supremacy in America, George M. Frederickson, writes that Nott's sentiments "would seem to belong at least as much to the history of proslavery and racist propaganda as to the history of science." Nott's biographer Reginald Horsman says bluntly that Nott's constant chest-thumping about scientific method notwithstanding, "there was no scientific method" in his essays on race or hybridity at all. Another recent authority, C. Loring Brace, writes that "Nott was less interested in the science of anthropology per se than he was in using selected portions of it to justify the continuation of special privileges for the white population in the antebellum American South. At bottom," Brace concludes, "Nott was a prototypical Southern racist."[79]

Nott was certainly a Southern racist, but in a probing and nuanced account, Terence D. Keel, a professor of genetics at UCLA, argues that he was not prototypical and, more provocatively, that he did not conduct his science in bad faith. In Keel's reading, Nott's unsparing focus on efforts by scientists to accommodate their theories of human origins to the Biblical account exposed a weak link in the then-dominant chain of scientific reasoning. Nott's polygenism, Keel argues, was actually "a valid scientific position, despite the obvious political consequences of the theory. Indeed," he continues, "the theories of the American polygenists need to be reassessed as legitimate scientific positions for their time and not merely window-dressing for racial politics."[80]

This may be, but to Nott's scientific arguments must be added his nonscientific eliminationist vision of a future in which inferior races either peter out genetically through a proliferation of infertile mulattoes or are exterminated by various means so that eventually all groups are back where God originally put them. Nott was a proslavery racist only in a restricted sense. He saw slavery not as part of God's plan but as a temporary makeshift, the only workable arrangement in the current state of confusion in which different species find themselves commingled through the accidents of history. A romantic concerning the chivalric Southern way of life, Nott thought himself coldly realistic in his understanding of the practice that made it possible. According to a British correspondent who visited him in 1861, Nott professed to have a "purely materialistic view of the question of slavery," and felt that "questions of morals and ethics, pertaining to its consideration, might be referred to the cubic capacity of the human cranium." He detested slavery, Nott told his guest, but could not imagine what would be done with the slaves if the institution came to an end.[81] With the economy and social organization of the South entirely dependent on slavery, and the number of slaves doubling every thirty years (partly as a result of white males momentarily setting aside their natural and honorable repugnance for black females and engaging in "ultra-ecclesiastical amalgamations"), the prospect of sudden abolition threatened social chaos and economic ruin, an immediate disaster for

Caucasians and Negroes alike. But even if the abolitionists were defeated, the example of Egypt demonstrated that slavery itself actually conduced to its own slow abolition as the superior race mixed its blood with the inferior so that in the end, the distinction, and the civilization, collapsed.

These were not prototypical anti-abolitionist sentiments. Nott's understanding of slavery was based on authorities in science and religion; he never defended the justice of the institution. As Stanton points out, Nott "had no love for slavery. He did not defend it with any of the numerous arguments then fashionable" and never pretended that it had been ordained by God.[82] Nor was this Nott's only deviation from orthodoxy. It is not known how many slave owners read *The New Orleans Surgical and Medical Journal*, but if any had, they might have been astonished to read Nott's cosmopolitan assertion that

> what we term the Caucasian race, is not a unit; it is, on the contrary, an amalgamation of an infinite number of primitive stocks, of different instincts, temperaments, mental and physical characters, etc. Egyptians, Jews, Arabs, Teutons, Celts, Slavonians, Pelasgians, Romans, Iberians, etc., etc., are all mingled in blood; and it is impossible now to go back and unravel this heterogeneous mass, and say precisely what each originally was.

However disorderly the origins, Nott argued, contemporary Caucasians were a species blended to a nicety and needed no further additions, unlike, for example, the Negro, for whom "a small trace of white blood . . . improves his intelligence and moral character"—a striking statement given Nott's arguments about the natural antagonism between the races, but less so given Nott's deep experience with plantation life.[83]

Unable to see beyond the current arrangement of the world around him, Nott fixed a distant gaze on the world that was at the beginning and the world that is to come, in which Caucasians would fulfill their divinely ordained destiny and Negroes would disappear everywhere except Africa. In that eventuality, regardless of how it came about, the statistics on Negro mortality would be beyond calculation. Insurance companies be warned: the value of colored life outside Africa would approach zero.

Given this vision, Gliddon's mummies and Morton's skulls—eventually over eleven hundred of them—appear less like precious sources of empirical evidence than as death's-head obsessions of those who, whatever their politics or morality, saw the goal of history as the perfection of their own race achieved through the elimination of millions of people who impeded progress toward that perfection. Nott was silent on the radical steps required to realize this goal, closing one of his lectures with customary humility, saying that "wiser heads than mine" must determine them in accordance with "the laws of God, written in the Book of Nature."[84]

Types of Mankind *and the Making of Americans*

By the end of his life the cheerful, hopeful, earnest, dedicated, and pacific Morton had effectively been conscripted into the service of scientific positions, social arguments, and historical understandings he had once repudiated. Created, driven, and controlled by Nott, the American School had claimed Morton as its spiritual and scientific leader. And he was: with the assistance of Gliddon, Morton had slid by degrees from positions for which he had developed some empirical evidence to positions for which no such evidence could even be imagined. From the Egyptian mummies, paintings, and other artifacts discussed by Morton in *Crania Americana* and *Crania Ægyptiaca*, conclusions had been drawn that could easily seem to legitimate the transatlantic slave trade and ultimately the extermination of all races except one in all places other than the African equatorial zones. With what surely seemed to him the best of intentions, Morton had put himself in the position of the scientific authority behind what would become the American School and the key transitional figure linking Gliddon's Egyptology and Nott's brutal vision of the earthly Caucasian paradise. A private autopsy on Morton revealed, with unintended suggestiveness, a "large but flabby heart."[85] With greater culpability than Blumenbach, whose affinity with Meiners was more imputed than real, Morton permitted himself to become a Northern, antislavery version of Nott so that one ran into the other with no clear boundary line between them.

In one respect, Morton was not alone. Some principle of gliding or elision, a drama of half-resisted seduction, seems to have been at work in the formation of the American School such that the most deep-laid or consequential differences among the various dignitaries became erased or blurred, with conscious opposition becoming in some cases unwilled or half-witting collaboration.

Take for example the contribution of Combe to *Crania Americana*. In his "Memoir" of Morton in *Types of Mankind*, Patterson suggested that when Morton agreed to Combe's suggestion that Combe contribute an essay to his book, he must have been unaware of the disrespect into which phrenology had fallen over the preceding twenty years in the scientific community, which had come to view the practice as "the property of itinerant charlatans, describing characters for twenty-five cents a head."[86] Out

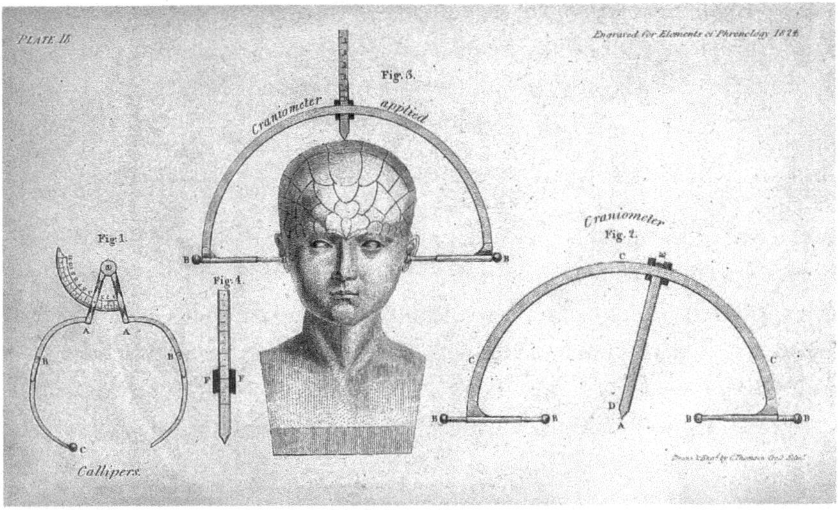

FIGURE 3.11. From George Combe, *Elements of Phrenology* (1824). Public domain. Source: Wellcome Collection, London.

of a spirit of diversity and inclusion, Morton had compromised himself and his work; as Stanton wrote in 1960, the participation of Combe "no doubt contributed to bringing the whole work into the same oblivion into which phrenology itself descended."[87]

But Combe himself had reason to feel conscripted for a cause that had no use for him. In a prefatory note to the volume, Morton declined to endorse the practice of phrenology beyond the most basic premises, that the brain is the organ of the mind and parts of the brain have different functions. To Combe's irritation, Morton made no use of phrenology in his descriptions of the various groups and even contradicted Combe's contention that brain size was correlated with intelligence by praising the ancient civilization of the (small-headed) Peruvians. So uncommitted was Morton's view of phrenology that one colleague actually wrote to congratulate Morton for having produced in *Crania Americana* "the strongest argument" against the practice.[88]

Combe had post-publication issues as well. As Morton's affiliation with Nott grew stronger and *Crania Americana* became associated with the polemics of the antebellum South, irritation might have become horror. A Scottish liberal, Combe was on record saying that American slavery was "a canker in the moral constitution of the country, that must produce evil continually."[89] And when Nott used Morton's work in support of his argument for deep differences among races, Combe may have felt beslimed by the association with Nott, for among Combe's deepest scientific convictions were the beliefs that all humans were one species, and that education and opportunity could literally grow the brain. If the Negro was free, Combe wrote in an essay describing his visit to the United States, his brain would enjoy "wholesome exercise" and the organs in his brain would increase in size.[90]

Their disagreements notwithstanding, Combe and Morton were effectively conjoined with the publication of *Crania Americana*. Then, with *Types of Mankind* presented as an *hommage* to Morton, they became collaborators in a joint project with Nott, signatories to a suite of arguments that neither would have accepted, at least not with the monotone zeal of Nott, about Caucasian supremacy, the fixity of races, the species difference between Caucasian and Negro, the natural antagonism between races, and the desirability of a Caucasian future in which all inferior races would be disappeared or quarantined in their allotted domains. Nor—their reservations, demurrals, and principled oppositions notwithstanding—could either Morton or Combe entirely disavow the association. As we have seen, Morton had in the years before his death in 1851 edged closer and closer to many of Nott's positions and had explicitly rejected none. And as for Combe—well, he had advanced as a scientific fact based on professional skullwork the proposition that "*permanent subjection to a foreign yoke*, is the result of an inferior aggregate development of brain, animal, moral and intellectual, in the people subdued, to that possessed by the conquering tribe."[91]

When he read these words in Combe's appendix to *Crania Americana*, Nott understood that he had found a phrenological friend, and included in his 1844 Mobile lectures the assurance that "no one doubts that the brain is the organ of intellect and instinct, and that the general facts of Phrenology are true."[92] The science of phrenology could explain why some peoples—Combe had cited Peruvians and Mexicans—were easily subjugated while others such as American Indians were not; to the group of the easily enslaved, Nott added Negroes, claiming phrenological and thus craniological and thus scientific support for a conclusion that the experience of a lifetime spent among the enslaved Negroes in the South had persuaded him was in any event a simple fact.

Like America, the American School was constructed through the absorption of disparate elements in an *e pluribus unum* spirit of assimilation. Nott was extraordinarily persistent and effective in enlisting or claiming support not only from those who did not protest their cooptation but from those who explicitly opposed or despised him. He took particular pleasure in exposing the grudging concessions and admissions made by the esteemed Prichard over the course of multiple editions of Prichard's *Researches into the Physical History of Mankind.*[93]

A passionate and lifelong advocate for the unity of humankind and a serious student of African ethnography and linguistics, Prichard had always positioned himself as the righteous foe of arguments like those being made by Nott. In his first edition in 1813, he had suggested that humanity, while diverse in various ways, had issued from a single stock created by God and originating in Africa, a boldly provocative assertion that implied that Adam was a Negro. But with each successive edition, Prichard, who was absorbing all available information about the peoples of the world as it was being accumulated and reported, seemed to become less committed to his own positions. The "out of Africa" assertion was an early casualty; in subsequent editions, Prichard would say only that the "natural and original complexion of the human species" was "melanocomous," or black—black *haired,* that is, a trait shared by indigenous people in America and Africa as well as by Hindus, Spaniards, Persians, and "black-haired Europeans in general."[94] When Nott saw Prichard relocating original blackness from skin to hair, he sensed a gentle retreat he knew he could exploit.

Casualties mounted with each edition of *Researches,* as Prichard's extensive ethnographic descriptions of the various populations of the world began to include harsh judgments, often reflective of French sources, of the primitivism or savagery of indigenous cultures. The Samoiedes, he says in the 1826 edition, "are a race of miserable, destitute savages" who "are in every respect in the lowest state of wretchedness and degradation of which human nature is capable."[95] The third edition was illustrated with ethnic "types" and skulls. By the final edition, words such as "repulsive," "stupid," "squalid," "wretched," and "ugly" were appearing with increasing regularity. The "generally ugly" women of Papua were described as "formed for servitude and obedience"; while the "Australian women, still more ugly than the men, have squalid and disgusting forms . . . the distance which separates them from the *beau idéal* appears immense in the eyes of an European" (1841, I, 246, 252, 257).

In the same volume, Prichard cited a 1785 study on the ratio of the brain to the central nervous system that concluded that "the Negro displays in this respect an approximation to the lower animals." And—undoubtedly very much to the approval of Nott, who had said in his article on insurance rates that it was merely human nature for the slave owner to treat an "unsound Negro" as he would a "superannuated horse"—Prichard also noted that the Bushmen of South Africa ("the most degraded and miserable of all humans") form "a transition from the genus homo to the genera of orangs and gibbons." The detailed account of the "Psychical Characters of the Bushman or Hottentot race" that follows is Dantean in its accelerating sense of disgust and horror.[96]

While Prichard never fully abandoned his monogenetic and humanitarian commitments—indeed, in 1843 he published *The Natural History of Man*, in which he affirmed once again his belief in the intellectual and moral unity of humankind—his later work gave increasing voice to the suspicion that race was a category comparable to species, and that the species line had been crossed somewhere in the vicinity of the Bushmen. As the champion of diversity, Nott was cruelly attentive to Prichard's waverings and soft collapses, and gleefully noted the shifting tone and modified arguments of each succeeding edition. "We behold him, year after year," Nott wrote, "like a bound giant, struggling with increasing strength against the cords which cramp him. . . . He set out, in youth, by distorting history and science to suit the theological notions of the day"—but "Alas! For his fame, Dr. Prichard continued to change his costume with the fashion."[97]

To Nott, Prichard was another Morton, a man honest enough to note the facts piling up against his prejudices but not brave enough to abandon those prejudices altogether. In the case of Morton, Nott could take incremental concessions as evidence that Morton was moving in the right direction, implying that if he had lived just a bit longer, similarity would have become identity. In the case of the recently-deceased Prichard, Nott could mention the facts and mock the cowardice, presenting in a brutal four-page section at the very beginning of his "Introduction" to *Types of Mankind* his own positions as the logical terminus of Prichard's wobbling but inexorable course of self-modification.[98]

Henry S. Patterson, the modest, careful, scholarly author of the "Memoir" of Morton placed at the beginning of *Types of Mankind*, could be counted another half-willing, half-witting conscript to Nott's cause. By depicting Morton as a man of great "magnetic power by which he attracted and bound men to him," Patterson gave Nott an invitation to represent himself as one of Morton's grateful devotees.[99] In fact, as we have seen, over the course of the 1840s Morton had increasingly taken his bearings from Nott, but if Patterson had said as much, the scientific and moral authority of the American School would have been greatly diminished and the dedication of *Types of Mankind* to Morton would have been inexplicable.

Patterson's greatest contribution to the American School, however, was his stout defense of Morton against the charge that by agreeing to "*unalterable* diversities," he was endorsing the practice of slavery. Such "ethical consequences," Patterson protests, "appear to me not to be fairly involved" in the scientific question. Science can be—must be!—an autonomous and limited field, and surely one can believe in equal rights for all even while acknowledging the diversity of types. Does anyone truly believe that science can prove or disprove "the inalienable right of man to equal liberty with his fellows," or that "an argument in favor of slavery, true or false, may be based upon" a scientific inquiry into human variation?[100] While Nott could hardly respect Patterson's generous sentiments concerning inalienable liberties, he undoubtedly welcomed the support of those in the North and in Europe who—like him—disapproved in principle of slavery, for such support dramatically increased his legion of admirers by giving them a way to endorse his science with a clean conscience.

With Prichard revealed as a distinguished English scholar who reluctantly acknowledged the evidence for plural creation, and Morton embraced as one of the most admired scientists in the world who effectively conceded the question of a species distinction among races, and Combe honored as a leading Scottish scientist and humanitarian who provided hard evidence for Caucasian superiority, and Patterson welcomed into the fold as a disinterested party who denied any link between science and slavery, the American School had become a big tent, with Nott its impresario. No matter where they had originated, all roads led to Mobile, Alabama.

This impression was only confirmed by the inclusion in *Types of Mankind* of an essay by the brilliant Swiss naturalist and polymath Louis Agassiz. Like Morton, Agassiz represented another trophy for Nott, for Agassiz's endorsement represented a conversion from an earlier and quite explicit commitment to the principle of human unity.[101]

What changed him?

Apparently it was the service he received at a Philadelphia restaurant, where he had gone to visit Morton in 1846 shortly after arriving in America. Never having beheld a Negro in Neuchâtel, he was traumatized by the appearance and deportment of the waitstaff, and, fatally for his subsequent reputation, committed his reaction to paper in a letter to his mother (who out of maternal pride preserved this note from her gifted son) in which he described the appalling sight of their skin, their limbs, their hands, their lips, and their woolly heads. It was all he could do, he told her, not to bolt from the room rather than endure the proximity of "this degraded and degenerate race" who were surely "not of the same blood as we are."[102] Rattled by the experience and impressed by Morton, Agassiz

FIGURE 3.12. Louis Agassiz. Public domain. Source: Wikimedia Commons.

became receptive to the idea of the unalterability of racial characteristics. The following year, attending a conference in Charleston, South Carolina on the subject of the unity or diversity of the human race, Agassiz rose following a speech by Nott to declare his beliefs in polygenesis and a species difference between Caucasians and Negroes; showing the effects of recent exposure to Morton, he adduced evidence from the ancient monuments and records of Egypt in support of that difference.[103]

Nott's response from the podium, if any, was not recorded, but it may have been mixed. Gratified by the acceptance of his authority on the question of species difference, Nott might have been amused or even appalled by Agassiz's visceral contempt, which was so different from his own more theorized but also more experienced and familiar (if no less ungenerous, unimaginative, and false) view of the potentialities of Negroes. In Agassiz, Nott was confronting a caricature of himself, a man who took Morton's "theory of *repugnance*" entirely too personally.

In spring 1853, Agassiz visited Nott in Mobile and agreed to contribute to the forthcoming *Types of Mankind* an essay on the geographical distribution of animals and humans. Intended as a display of hyper-scrupulous scientific reasoning, the result has been described by a recent scholar as "the most embarrassing memorial to the career of that otherwise distinguished Swiss American naturalist."[104] Beginning as a learned discourse on the distribution of the fauna of the world into "distinct zoological provinces," the essay became, at the end, a solid contribution to Nott's American School. Citing Morton's definition of species as "primordial organic forms," Agassiz applied the principle of distinctness to human beings. "I am," he announced, "prepared to show that the differences existing between the races of men are of the same kind as the differences observed between the different families, genera, and species of monkeys or other animals . . . nay, the differences between distinct races are often greater than those distinguishing species of animals one from the other."[105] The line demarcating Agassiz from Nott had been as effectively smudged as the line separating Negroes from primates.

As to the origin of racial differences, Agassiz was even more obliging to Nott, for he essentially reproduced Nott's theory in *Two Lectures* that original stocks were placed by God in the climates best suited to them. "We must acknowledge," he said, "that the diversity among animals is a fact determined by the will of the Creator, and their geographical distribution part of the general plan which unites all organized beings into one great organic conception: whence it follows that what are called human races . . . are distinct primordial forms of the type of man."[106] This was Nott in a nutshell, and if it is not, as Brace argues, an explicit "justification" for slavery, it is certainly not an argument against it.[107]

By showcasing Agassiz at the beginning of *Types of Mankind*, Nott was inviting his readers to admire the world-conquering triumph of the American School. With the rapid growth of Agassiz's public reputation, this endorsement became increasingly valuable. As Brace says of Agassiz, "to the American public, his pronouncements were synonymous with the findings of 'science.'"[108] But in this case, it was science as Nott had defined it, homegrown American science with specific implications for American society. Indeed, the title of the chapter in which Brace discusses his contribution is "Louis Agassiz, Archetypal American."

Nott was not finished conquering the world. Science had capitulated, but history remained. Shortly after the publication of *Types of Mankind*, the indefatigable Nott arranged for a translation of the first book of Count Joseph Arthur de Gobineau's six-book *Essai sur l'inégalité des races humaines*.[109] He had read, or skimmed, that single text with great enthusiasm, feeling, not for the first time, that he had found a kindred spirit in a thinker whose convictions overlapped with his own and whose differences he could forgive and forget; indeed, he could ignore and suppress them.

The shared convictions were substantial. Like Nott, Gobineau believed in the inequality of the races, which he ranked from high to low, with whites constituting a racial aristocracy and darker-skinned peoples, especially those from Oceania, representing an almost inconceivable level of degradation. Gobineau believed, too, that the characteristics of any given civilization reflected the capacities of the dominant race. Like Nott, Gobineau held that racial differences were so profound as to be unalterable except through the "crossing of blood," a form of corruption that in the case of Caucasians and Negroes represented an affront to nature and a disaster for civilization. Gobineau also shared Nott's conviction that racial mongrelization had advanced to such an extent over the long history of humankind that only a tiny number of peoples had managed to resist or somehow neutralize the pollution of their racial stock. And lastly, both men shared an appreciation for Morton, with Gobineau even reprinting Morton's footnote recording the statistics for the cranial capacities of the major races.

As to the differences with Gobineau, if Nott had enumerated—or fully comprehended—them, they would make a far longer list. Obsessed with the Middle Ages, Gobineau was unconvinced by and uninterested in studies of hybridity, the definition of species, or empirical science generally. Gobineau's Chapter 10, "Question of Unity or Plurality of Species," actually rebuts Nott's arguments about hybridity point by point. Conventionally religious, Gobineau concedes the limitations of the Mosaic text but defends the monogenetic unity of the human species in the first dawn of humanity. At some point, Gobineau believed, differences had arisen, but he had no interest in the questions of how or why; for him, the only relevant point was that they were now fixed. While he held that the number of what he called "primal races" was small—he identified three—he gave the name of race to many other groupings and argued that the hidden cause of wars, revolutions, or rebellions was interracial conflict. The French Revolution, to take the most dramatic recent example, could be understood as the long-deferred revenge of the inferior Celts and Gauls against the Teutonic "Frankish" aristocracy that had conquered them in the fifth century AD.

As these examples suggest, Gobineau was as disdainful toward the masses in any race as he was toward races he considered inferior. Gobineau often used race and class almost interchangeably, so that classes took on a racial character and races ranged themselves in classes.

His deepest antipathy was not toward inferior races, which were not truly responsible for their limitations, but toward democracy. As Arendt said, Gobineau "invented racism almost by accident" in the course of his wide-ranging ruminations on superior and inferior groups.[110] Deeply pessimistic, Gobineau declared bluntly that mankind was incapable of infinite progress and the white race was currently in the process of annihilating itself.

Among the very few exceptions to this principle of inevitable racial suicide were the Jews, who almost alone among the white race had long resisted such corruption. But even they were not direct descendants of the pristine fountainhead of humanity, which had long since vanished. This was the ancient "Aryan" race, which Gobineau depicted as a powerful, beautiful, and relentlessly creative group that had, in the inexorable course of their migrations, founded virtually all the great civilizations of history. As the Aryans spread themselves over the world, they mingled their blood with others. For Gobineau, such mingling was a tragic necessity. While civilization depended on such mixtures and racial purity was stagnation, the corruption of the primal stock had effected the extinction of the Aryans so utterly that their very existence could only be inferred from the *mélange* that now populated the earth.

None of this had an echo in Nott; and yet he sponsored and oversaw a translation of Gobineau's first volume, which appeared in 1856 with the title *The Moral and Intellectual Diversity of Races*, and commented in the appendix that he had "seldom perused a work which has afforded me so much pleasure and instruction."[111] He had in effect pleased and instructed himself, sponsoring the translation by Henry Hotz, a precocious young Swiss citizen then working as a tutor in Mobile. Hotz had arrived in the United States at the age of fifteen but had quickly demonstrated unusual and indeed almost inexplicable enthusiasm for Southern culture as exemplified by Nott. Favored with this assignment, Hotz worked quickly, providing in addition to the translation a one-hundred-page "Analytical Introduction" and a great many almost invasive historical notes, many of which took editorial exception to Gobineau's arguments. Nott's appendix added to the sum of knowledge about crania by reporting on statistics provided by various hatmakers with whom he had corresponded.

The introduction itself is an independent contribution to science. Displaying a striking sense of his own importance, Hotz, who had not yet attained voting age in his adopted country, begins by weighing in on the numerous questions and issues that were dominating the scientific discourse on race as well as on some larger questions concerning humanity in general, venturing criticisms of Prichard, Adam Smith, Alexis de

FIGURE 3.13. Henry Hotz. Public domain. Source: *Encyclopedia of Alabama.*

Tocqueville, John Locke, and many others. He volunteers extended personal ruminations on the best forms of government, the meaning of civilization, and the intentions of God. He rehearses with considerable feeling the history of "the white and the black races, supposed by many to represent the two extremes of the scale of gradation. The whole history of the former," he says, "shows an uninterrupted progress; that of the latter, monotonous stagnation. To the one, mankind owes the most valuable discoveries in the domain of thought, and their practical application; to the other, it owes nothing."[112] As evidence of the scientific consensus on these issues, Hotz reproduces a chart from Robert Gordon Latham's *Natural History of the Varieties of Man* (1850) in which each of the three major races is evaluated for qualities of "intellect," "animal propensities," and "moral manifestations," and graded on a range from "feeble" to "very strong," with unsurprising results.

Most remarkably, Hotz announces that he, the translator, has a definite "purpose"—the restoration of a traditional understanding of "nation" to mean not just a people united by consanguinity and a particular form of government but rather "*a population consisting of homogeneous ethnical elements.*" Hotz proposes, in other words, to use his translation, which he

	BLACK RACES, OR ATLANTIDÆ.	YELLOW RACES, OR MONGOLIDÆ.	WHITE RACES, OR JAPETIDÆ.
INTELLECT	Feeble	Mediocre	Vigorous.
ANIMAL PROPENSITIES	Very strong	Moderate	Strong.
MORAL MANIFESTATIONS	Partially latent	Comparatively developed	Highly cultivated.

FIGURE 3.14. From Robert Gordon Latham, *Natural History of the Varieties of Man* (1850); reproduced in Henry Hotz, "Analytical Introduction," in *Moral and Intellectual Diversity of Races*, by Count A. de Gobineau (1856), 94. Public domain. Source: Internet Archive.

freely admits is not "*literal*," to redefine the nation as a race.[113] The "ethnical elements" that define the Hotzian nation do not, however, consist of physical traits alone but also include certain political practices and cultural values, including respect for women, love of liberty, representative government, trial by jury, "and all the discoveries in political science upon which we pride ourselves most."[114] This statement is included in a twelve-page "Introductory Note" preceding Gobineau's chapters 8 and 9.

Such an assertion of liberal and democratic principles, especially when attached to a polity, the slave-holding states of the American South, that would soon rise up in rebellion, represented an impudent rejection of Gobineau's most cherished convictions, one of which was that the population of the United States was not an ethnically homogeneous people at all, but "a very mixed assortment of the most degenerate races of olden-day Europe . . . the human flotsam of all ages. . . . [I]t is quite unimaginable that anything could result from such horrible confusion."[115] Since, however, this passage appeared in Gobineau's second volume, and Hotz had translated only the first, and not all of that, these reflections were not included in his book; it is entirely possible that neither Nott nor Hotz had "perused" them at all.

Still, even Book I provided challenges, and some ingenuity was required for Hotz to claim that the American nation could be included in Gobineau's phrase "our civilization." This was the task of that "Introductory Note," which preceded Gobineau's discussion of "civilization." In it, Hotz explained that the core group of the American nation, having emigrated from northern Europe and particularly from England, "presents a happy fusion of some of the most distinguished branches of the German family."

> If we now glance at the United States, we shall there find—at least in the first years of her national existence—a pendant to what has been asserted of England. The elements of the population of the original thirteen States, were almost exclusively of English, Lowland Scotch, Dutch, and Swedish blood; that is to say, decidedly Germanic. Ireland was as yet slightly represented. France had made but inconsiderable contributions to the population. Since we have assumed a rank among the great powers of the earth, every portion of the inhabited globe has sent us its contingent of blood, yet even now, the great body of the nation belongs to the Teutonic race.[116]

In (the untranslated) Book VI, Gobineau had commented precisely on the American insistence on an "Anglo-Saxon" core, saying that "this name of Anglo-Saxons seems to flatter the imagination of the inhabitants of the great transatlantic confederation, in spite of the increasingly equivocal right that the present population may have to claim it."[117]

Gobineau deplored random mongrelization particularly between black and white, but in an argument that must have struck Nott and Hotz as disastrously misconceived, he reluctantly conceded the benefit—and even argued for the necessity—of such a mixture for the emergence of artistic genius, for which, he noted, "the black element is certainly indispensable . . . for we have seen what outbursts of fire, flames, sparks, vivacity and spontaneity are intrinsic to its soul."[118] "I do not deny," Gobineau wrote at the conclusion of Book I, that the worlds of art and literature have benefited immensely from such mixtures, which have produced "wonders for which we must needs be thankful" [*de merveilles auxquelles il faut applauder*].[119]

This was so spectacularly off-message that Hotz declined to translate it at all, and even interpolated refutations of such misguided generosity in the "Introductory Note" and elsewhere. "Much has been said," he wrote (declining to specify who said it),

> of the effects of ethnical mixture. Many consider it as decidedly beneficial, others as decidedly deleterious. It seems to me susceptible of mathematical demonstration, that when a very inferior race amalgamates with one of higher order, the compound—though superior to the one, must be inferior to the other. In that case, therefore, mixture is injurious. But when various branches of the same race, or nearly cognate races mix, as in the case of the Saxons, Angles, Danes, and Normans, the mixture cannot but be beneficial. For, while none of the higher qualities are lost, the compound presents a felicitous combination of some of the virtues peculiar to each.[120]

Not surprisingly, Gobineau was appalled, feeling himself to be more traduced than translated by a work that in addition to its shocking freedom of adaptation, reprehensible advocacy for democracy and polygenesis, ridiculous claims of mulatto infertility, and absurd and unwarranted American racial pride, had ignored his last five books altogether, which had included vast amounts of commentary on the history of civilization that supported his generalizations as well as the damning criticisms he had made of the mongrelized population of the United States and the brutality of the Anglo-Saxon's treatment of Indians and slaves.[121]

The English translation was, then, not a mere re-presentation of Gobineau's text in English but a project with a particular and local purpose: to provide scholarly, European, and aristocratic support for arguments about the inequality of the races and the inferiority of the Negro in particular that were favored by Nott and others in the increasingly secessionist South. For reasons that remain mysterious, Hotz embraced this purpose wholeheartedly, acquiring, if he did not already have, the attitudes and assumptions about race that were common in the American South at the time. He embraced these so fervently that he volunteered to serve during the Civil War, at first in a company led by one of Nott's sons, and then as a pro-Confederate propagandist in England. A committed supporter of slavery, he would, at the end of the war, neither give up his American citizenship, which he acquired one day after his twenty-first birthday, nor return to a post-slavery America. He remained in Europe after the war and died in 1887.

Nott discovered in Gobineau a highly congenial and adaptable mass of argumentation conducing to the conclusion that races were as unequal in endowment and capacity as they were different in physical appearance. Gobineau's various other arguments about the ethnic character of the French aristocracy, the necessity of restoring traditional hierarchies, the splendor of the chivalric era, the degeneracy represented by the practice of slavery (a constant theme), the brilliance of the Aryan race, or the dangers of democracy had little value or interest for Nott. Neither he nor the loquacious Hotz so much as mentioned Aryans. And yet, in the corrupted form in which Gobineau was presented in a translation that remained the only one available until 1915 when Adrian Collins produced a more accurate version—still, only the first of six books—Gobineau became a fellow-traveler in the American School along with Prichard, Morton, Gliddon, Combe, Patterson, Agassiz, and Hotz. A mixed group to be sure, but not without distinction, and all of them gathered into the magnanimous and forgiving embrace of Nott.

"A MAN NOW"

The ideological and scientific affiliate of the losing side in a massively destructive civil war, the American School might seem a sad footnote in the history of science. But while the intellectual prestige of the American School ended at Appomattox, the issues raised by *Types of Mankind* and other publications remained. The implications of "diversity," the relationship between race and species, the significance of hybridity studies, the information to be gained from craniometry, and the relative positions of the various races in the order of nature—all these questions, detached from the immediate issue of slavery and the American setting, took on a new life in the scientific discourse of Europe. This new life will be the subject of the next chapter, but in the United States the fundamental question driving all these secondary inquiries was paramount because it spoke to the identity of the nation both before and after the war. The question was, In a mongrel world, could some humans have a stronger claim to humanity—human reason, human rights, human dignity—than others?

What is a human being?

This question was already greatly on the mind of one of the earliest and most engaged readers of *Types of Mankind*. Thirty-six years old when the book appeared, Frederick Douglass may well have felt that Nott was in mysterious ways speaking directly to and about him. Born a slave to what Nott would describe as an "unfeeling master," Douglass, as the presumptive son of that master, was also a mulatto of the sort that Nott described as a degenerate hybrid.

There is no evidence that Douglass ever felt that he was comprised of two distinct species, or that he, or half of him, stood at the perilous margin between humanity and primates. Nor, as the father of five children, did he have any reason to think that he was infertile or that he represented a station on the path to biological extinction. In the fifteen years since he had escaped slavery, he had educated and trained himself so that he was a commanding presence in all companies, quite the opposite of the figure of the mulatto, feeble in mind and body, posited by Nott's theory of racial mixture. In short, as he took the podium on July 12, 1854, at Western Reserve University (now Case Western Reserve) to address the Philozetian Society on the subject of "The Claims of the Negro, Ethnologically Considered," Douglass had every reason to feel that he was an incorporate refutation of that theory and of the American School generally.[122]

Douglass was at once the most compelling voice and the strongest argument for the abolition of slavery. A self-taught student of the Bible,

FIGURE 3.15. Frederick Douglass, from *My Bondage and My Freedom* (1855). Public domain. Source: Wikimedia Commons.

he had mastered a rhetoric of righteous indignation, prophetic warnings, and moral denunciation that focused and fueled an abolitionist mood of increasing determination. On this occasion, he was directing his ferocity at two targets in particular, the Fugitive Slave Act of 1850 that provided for the capture and return of runaway slaves, and the Kansas-Nebraska Act of 1854 that, by allowing for popular sovereignty in those two territories, had created the possibility that the number of slave states would increase. Just a few weeks earlier, Douglass had given a speech denouncing the Kansas-Nebraska Act as a "hell-black scheme for extending slavery" and declaring that "the time for action has come."[123]

In this extraordinarily fraught political and moral context, ethnology might have seemed a curious choice for a public address. Indeed, so remote from Douglass's—and the nation's—immediate concerns was ethnology that it appears even to some of Douglass's most sympathetic readers that this address simply has no place in the great arc of his career. Neither Philip Foner's nor David Blight's (nine-hundred page) biography mentions it at all.[124]

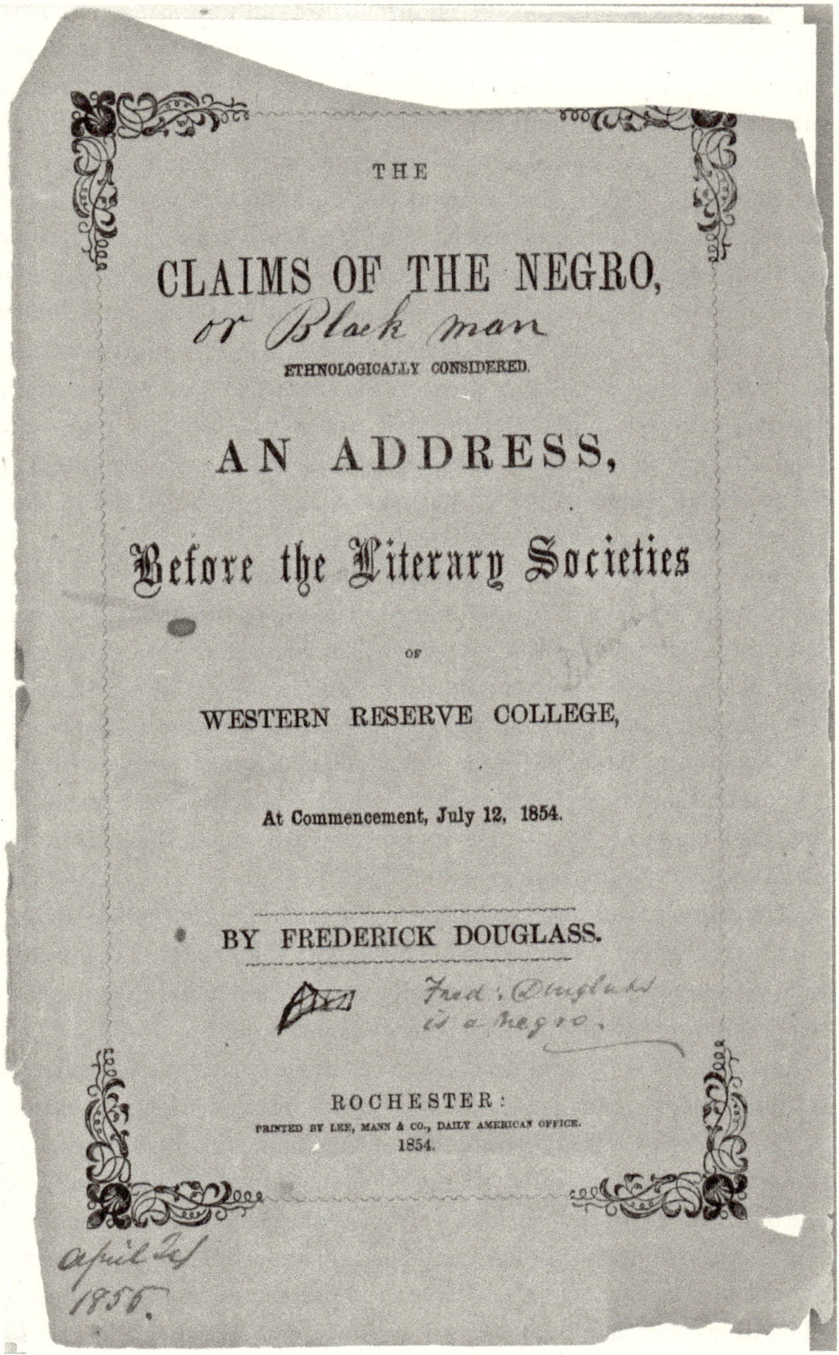

THE

CLAIMS OF THE NEGRO,

ETHNOLOGICALLY CONSIDERED.

AN ADDRESS,

Before the Literary Societies

OF

WESTERN RESERVE COLLEGE,

At Commencement, July 12, 1854.

BY FREDERICK DOUGLASS.

ROCHESTER:

PRINTED BY LEE, MANN & CO., DAILY AMERICAN OFFICE.

1854.

FIGURE 3.16. Frederick Douglass, "The Claims of the Negro, Ethnologically Considered" (1854). Public domain. Source: Library of Congress, Washington, DC.

If Douglass is to be considered primarily as an impassioned activist, the Western Reserve speech appears to be a relatively trivial and even distracting production. But if he is seen as an intellectual determined to identify not only immediate urgencies but the deeper sources of injustice, the speech can be seen as central and even essential to his larger project. For in it, Douglass, acutely aware that the address represented "a new chapter in my humble experience," actually thinks beyond slavery, beyond politics, beyond the present moment, and even beyond the American context. Of course, these were not far off since, as he ironically put it, "there is no doubt that Messrs. Nott, Gliddon, Morton, Smith, and Agassiz were duly consulted by our slavery-propagating statesmen" as they framed the Kansas-Nebraska bill.[125] But that bill, and all the other outrages that might have been identified, were, Douglass realized, mere symptoms of a way of understanding the world that was being promoted not by slave owners alone but also by respected scientists.

A new front in the "moral battlefield" had to be opened up in order to contest the "phalanx of learned men—speaking in the name of *science*" but promulgating "scientific moonshine" to advance concepts that were both false and foundational to the institution of slavery.[126] Shielded by the specialized vocabulary, formal protocols, and social prestige of science, the anti-abolitionist forces of the kind clustered in *Types of Mankind* argued not just that slavery was immensely profitable, which went without saying, but that it conformed to natural law, the details of which, Douglass noted, had also gone without saying.

The scientific authority conveyed by *Types of Mankind* on the subject of natural law had made it possible for people who regarded themselves as intellectually advanced and morally sensitive to extend the benefit of the doubt not only to a practice regarded by most of the world as a barbarous relic and a humanitarian disgrace but also to other less explicit or extreme practices of dispossession. By addressing the discipline of ethnology, then, Douglass was both attacking the enabling fiction of slavery and thinking far into the future with the understanding that slavery was only one of the many possible forms that "racial" oppression might take. If the Negro was to flourish in a post-emancipation society, Douglass seemed to understand, the struggle would have to be prosecuted on other fronts, in a language other than prophetic denunciation and apocalyptic warning.

In essence, the project of Douglass's essay is to identify, in the heterodox compendium that was *Types of Mankind*, a series of rooms, each with doors opening with decreasing levels of resistance onto the next. Nott's merely terminological contention that races were equivalent to species—an arguable case since neither term was clearly defined—opened a creaking door that led to the claim that races, like species, had fixed and distinct capacities that could be identified, described, and ranked. From there, a swinging door opened onto the assertion that, ethnologically considered, the Negro was not a man in the same sense as the Caucasian and belonged in a different category of creation altogether. From there, a half-open door revealed a room in which stood the stark conclusion that Negroes were natural slaves and Caucasians natural masters. Thus was the reader conducted from a clarification of terms to antebellum Alabama.

Rather than beginning at the first door with the terminological question, Douglass attacks the last, the conclusion that the Negro was a distinct species singularly suited to enslavement. His refutation of this conclusion is conducted in terms so elementary as to confirm for some his reputation for being far stronger in emotive rhetoric than logic or reasoning. The Negro, Douglass points out, has two hands like other humans, and can, like them, laugh and cry, and acquire and retain knowledge. The distinction, understood without instruction by all, between men and brutes is "as eternal as it is palpable"; and in short, by all the tests one may devise, "the negro is a MAN."[127]

Such a case would scarcely seem to need a Frederick Douglass to make it, and yet as Douglass notes, an article had recently appeared in the *Richmond Examiner* that, having raised in a spirit of scrupulous fairness the possibility that slavery might be a grievous wrong that unjustly deprived the slave of his rights, concluded by declaring that the Negro in fact has no such rights because, as Douglass paraphrased it, "HE IS NOT A MAN."[128] The contemporary case for the manhood, or personhood, of the Negro was not, apparently, self-evident to all.

Dispensing with Nott's conclusion, Douglass turns his attention to another, subtler source of error: the ethnological principle of diversity, which, he argues, corrupted the entire chain of reasoning on the subject of race. In a sense, he says, the principle of diversity is entailed by the discipline of ethnology, which is in turn entailed by the assumptions of science. Modernity produces race; or as Douglass puts it, "Science is favorable to division." Before the modern age of science, if you said that a particular being was a man, it would have been assumed that the being in question "had a common ancestry with the rest of mankind." Now, following the "profound discoveries in ethnological science" of "the Notts, the Gliddens [*sic*], the Agassiz, and Mortons," it has become possible to argue that this is not the case.[129] Science has produced rather than reduced confusion, for all human differences, according to Douglass, can be attributable to isolation, climate, and the conditions of life. The character of the Negro that conforms so nicely to the needs of the plantation has in effect been grown on the plantation with the other crops and would take different forms in different circumstances.

Diversity, Douglass implies, is a scientific name for the principle endorsed by Caesar, Machiavelli, Napoléon, and others faced with the task of sovereignty: divide and conquer.

To division Douglass opposes unity, and to science religion. The "unity of the human race," he says, is "all too plainly taught in the Bible to admit of cavil." The "Inspired Apostle" affirms this, writing that "God has made of one blood all nations of men for to dwell upon all the face of the earth." To insist on "diversity of origin" is not only "utterly subversive of the happiness of man [but] insulting to the wisdom of God."[130] Drawing expansive inferences from dubiously precise definitions of the squirrelly terms race and species, and positing categories and distinctions in nature for which there is no evidence, science misunderstands reality, makes of itself the enemy of the human family, traduces the Bible, and blasphemes against God.

Douglass argues not against a flawed concept of race but against science and its predisposition to division. In fact, he, like Blumenbach, was deeply suspicious of race. "The proposition to which I allude," he says in "The Claims of the Negro," "is that what are technically called the Negro race, are part of the human family, and are descended from a common ancestry, with the rest of mankind."[131] The only group Douglass was comfortable describing as a race was humanity at large and in toto.

Douglass's "ethnological claims" are not, then, based on the capacity of Negroes to rise to the same level as other groups, a capacity he both presumed and proved. They rest on the more fundamental fact that "what are technically called" races are mere variants and that "human rights stand upon a common basis."[132] In an 1884 interview, Douglass stated directly what his earlier writing and speaking had strongly implied, that the assumption that humankind consisted of biologically discrete groups of people was part of the problem. "Race pride," he told the interviewer, is "a positive evil" because "God Almighty made but one race."[133] And in an 1893 speech, "Abraham Lincoln—The Great Man of Our Century," he said, "I feel it more to be a man and a member of the great human family than to be a member of any one of the many varieties of the human race, whether Anglo-Saxon, Anglo-African, or any other."[134] He looked forward to a time when the original and foundational unity of humankind would be recognized once again as the process of amalgamation made inevitable by coexistence produced increasing numbers of people such as himself.

With, as Douglass pointed out, one-seventh of all Americans being descended from Africans, the United States was clearly the place where this goal had the best chance of being achieved, at least with respect to Caucasians and Negroes. In his introduction to *My Bondage and My Freedom*, the black physician James McCune Smith calls Douglass a "Representative American man—a type of his countrymen"—and his book "an American book, for Americans."[135] A composite exemplar of a composite nation, and a man who had passed from servitude to a position of great public prominence, Douglass had a more comprehensive experience of American life than most citizens, and insofar as a different "American School" of thinking has arisen to rival and counter that of Morton, Nott, Gliddon, and Agassiz, Douglass may be considered its founder.

And yet Douglass did not ground his thinking in any special American dispensation; nor did he treat American slavery as anything more than a local practice of which he happened to have personal experience. The larger threat, he felt, came from a scientific way of thinking that had the capacity to outlive slavery and its manifest brutalities, implanting itself in the culture of modernity, where, as Nott and his confederates demonstrated, it could discover or invent various principles of division with a perfectly clean conscience, as if no moral choices were being made at all. With its emphasis on classification, ethnology itself underwrote the institution of slavery while undercutting what Douglass calls in *My Bondage and My Freedom* "the natural and inborn right of every member of the human family" to freedom.[136] A sense of human rights based on a common humanity had, he said there, been awakened by his boyhood encounter with *The Columbian Orator*, a collection of essays and dialogues with an abolitionist cast that was widely used to teach the art of public speaking. When he began to speak in public, he was thinking of human rather than racial rights, and injustice in general rather than the specific injustice of slavery.

This broadening of perspective is clearly visible in the difference between the account given of a physical struggle with his then-master Edward Covey in the first autobiography, *Narrative of the Life of Frederick Douglass, an American Slave* (1845) and the account given ten years later in *My Bondage and My Freedom* (1855).[137] In the first, Douglass describes how his two-hour combat produced a profound transformation in his self-understanding. "The battle with Mr. Covey was the turning-point in my career as a slave," he wrote; "It rekindled the few expiring embers of freedom and revived within me a sense of my own manhood." In the second version, which he began writing shortly after delivering "The Claims of the Negro," the event itself is rendered in greater detail, but the real difference is that the turning point is described not as an awakening of a source of personal strength or virility but as an assumption of full membership in the human community, a membership denied not just by Covey and his kind but—he now realized—by the science that identified him as a member of a discrete human group, a "Negro." "I was nothing before," he writes; "I was a MAN NOW."[138]

The fate of the American School represented by *Types of Mankind* was decided not only by the Civil War but also by the revolution in science represented by the theory of evolution. Darwin established a common human ancestry calculated not in the few thousand years contemplated by Morton and Nott but in the far longer spans required for *Homo sapiens* to differentiate itself from primates. Highly controversial when it was announced in 1859, the concept of evolution steadily gained adherents, its support coming at the expense of those such as Nott—and Douglass—who were still thinking about God-created categories. The Darwinian world of random, adaptation-driven change had no place either for God or for the concept of race as Nott understood it. Nott refused to concede the victory to Darwin, however, and in a pamphlet published in 1866 he roundly condemned all "developmental" accounts that represented humankind, including the noble Caucasian race, as "nothing more than a developed worm."[139] But by then the tide of battle had once more turned against him.

In the years following the Civil War, Douglass became increasingly recognized in the United States and abroad as the "venerable Moses of his race."[140] Unlike Moses, however, he took a position in the government, and was sent as minister and general consul to Haiti, charged with the responsibility for negotiating the creation of a US naval base at the Môle Saint-Nicolas. In that capacity, he found himself in late 1889 confronting the Haitian foreign minister, a man of exceptional force and presence named Anténor Firmin, who informed—and persuaded—Douglass that American ambitions in Haiti were politically, historically, and morally unacceptable.[141] Failing in his diplomatic mission, Douglass returned after ten months to a country in which lynchings, often justified by the charge of rape against black men, were becoming increasingly common. This was a subject on which Douglass, almost certainly the child of a raped black woman, had strong feelings, speaking about it many times during the last year of his life. He died in 1895, one year before the *Plessy v. Ferguson* case legitimated "separate but equal" facilities for black people.[142]

Surviving members of the discredited American School, on the other hand, found ways to flourish in a world newly receptive to their gifts. Agassiz became a renowned professor at Harvard, while Nott moved north and remade himself in New York City, abandoning ethnology and establishing a successful medical practice on West Twenty-Third Street. Overcoming complaints from some of his gynecology patients about his "rough manner and want of dexterity," he became president of the New York Obstetrical Society.[143] While the elaborate skein of arguments embalmed in *Types of Mankind*, decoupled from its political implications, died an unmourned death, Nott was able to reinvent himself in Northern professional society, a beneficiary of the eagerness of the North to bind up the nation's wounds and move ahead in a spirit of reconciliation. The defeat of the Confederacy severed the connection between race and slavery in the sense that one could think of the former dispassionately without being burdened or distracted by the latter. With the passage of time, it became clear that Nott's basic premise of deep and radical "racial" differences was widely shared, having become ingrained in the popular mind in both North and South.[144]

It is a fact well worth pondering that while, "to a later generation," Nott might seem to be "among the lunatic fringe of racist writers . . . he was never that. He was," according to his biographer, "controversial but he was in the mainstream of American, not just southern, ethnological writing."[145]

Nor was his eminence a merely local phenomenon. His work on hybridity, no longer seen as ancillary to proslavery arguments, was taken seriously by European scientists including the brilliant French physician Paul Broca, the founder of the Société d'Anthropologie de Paris, whose "Memoirs on Hybridity," a three-hundred-page collection of essays from 1858–60, cited Nott generously and respectfully, with particular attention to Nott's research on mulattoes.[146] When the internationally renowned Nott returned to Mobile just before his death in March 1873, he was greeted, in the words of one observer, with "one continued ovation."[147]

Nearly thirty years later, another loyal son of the South made a successful career in the North. A North Carolina native, Thomas Dixon developed a large congregation in New York City before launching a career as the most popular lecturer in the country, becoming a spokesman for liberal and humanitarian causes. For reasons that remain unclear, Dixon abruptly abandoned both public speaking and liberalism after the turn of the century and devoted himself to literature, writing novels about the South under Reconstruction. His extraordinarily successful *Clansman* trilogy (1902–7) was made into an ambitious film in which dramatic and even melodramatic expression was given to the premises of the American School: Negro inferiority, special contempt for and even dread of the mulatto, a heroic conception of Southern Caucasian culture, an obsessive fear of sexually voracious black men, and a high and righteous disdain for the misguided Northerners who questioned these premises.

That film, appearing in 1915 just as the nation was commemorating the fiftieth anniversary of the end of the Civil War, was given the title of *The Birth of a Nation*. Depicting the post-war founding of the Ku Klux Klan as a heroic and even chivalric response to the excesses of Reconstruction, the film was directly instrumental in the rise of the "second era" of the Klan, which launched a far more violent, widespread, and systematic assault on black people than the first, in the 1870s. Supported by the adoption of "Jim Crow" laws in thirty-six states, the ideology behind this wave of violence was, according to many historians, nevertheless effective in binding the North and South together in a shared memory of the great war that once divided them, even as it made colored lives less insurable than ever. An ideology that had driven a failed scientific project was repurposed as a perverse mechanism of national reconciliation, acquiring a name that seemed both a concession and a clenched fist of defiance, a name that might almost have served as an inscription on Nott's headstone: The Lost Cause.[148]

CHAPTER FOUR

Equal or Unequal?

BROCA / GOBINEAU / FIRMIN

Science and the Mission Civilisatrice

The youngest, least credentialed, and darkest person in the room, he was as outside as an outsider could be and still be on the inside. But that's where he was. A proud, contentious, and even vain man, he had always been resentful of any assumption of entitlement by others; and yet, as a newly invited member of this most prestigious scientific organization, he had to recognize that he himself could now be a legitimate target of such resentment.

He permitted himself the thought that his previous sensitivities had been unwarranted, reflecting a lack of confidence and self-esteem that was perhaps understandable given his chaotic recent history. From his current perspective, the world looked different; he himself looked different, even to himself, now that he was where he belonged—among his intellectual peers, inside the skull of the world as it were, where the deeper issues of human existence were being confronted and addressed. It was all a matter of perspective.

As he cast his eye around the room, the young man reflected that there was no place on earth where the human species was shown to better advantage. The Société d'Anthropologie de Paris was at once a *tableau vivant* of the most eminent scientists and scholars in France—in Europe, in the world—and a living monument to its recently-deceased founder. The great Paul Broca had earned his immense reputation by casting the bright light of modern science onto, or into, the darkest continent of all, penetrating the brain and locating the exact regions responsible for aphasia (the third circumvolution of the left frontal lobe) and for the production of speech (the ventroposterior region of the frontal lobes). A staggering scientific achievement. Now, in July 1884, the Society was carrying forward the project that had dominated Broca's last twenty years, the study of humanity through empirical science—anthropometry it was called, a branch of the still-new science of anthropology—in which the great man had made astonishing progress by developing practices of cranial measurement that were being emulated around the world. He had actually written the textbook![1] The royal road to human self-understanding ran, Broca believed, not through the soft, perishable, and variable tissue of the brain but through the durable, measurable, and comparable skull.

The official position of the Society was that the gradual perfection of instruments of calibration and the accumulation of data resulting from thousands and even millions of measurements would yield results that would illuminate the mystery of the age: the origin, nature, and implications of human variation. Precise measurements would replace traveler's tales, missionary accounts, and all too speculative studies of languages and migrations. Differences in skull shape and volume, which had been discussed in preliminary and sometimes, admittedly, embarrassingly unsophisticated ways for decades, would become objects of professional attention, producing hard knowledge about the physiology and even, perhaps, the intellectual and other capacities of human groups.

With this knowledge, science would come closer to understanding the specific kinds of differences among the peoples of the world. Since the nations of Europe were steadily expanding their interests and influence not to mention their territorial possessions, the consequences would be enormous. Among the beneficiaries would be science itself, which would acquire crucial information about issues currently dominated by politicians, moralists, generalists, and amateurs of various kinds. The scientists of the world could even formulate new ways of thinking about what were formerly theological or mythological issues of human origins. They could write a new Book of Genesis for the modern age!

Everyone in the Society understood the extraordinary importance of the undertaking and the sacrifices it would require. Currently, the main obstacle to progress was the lamentable paucity of skulls, especially skulls formerly in the possession (if that is the correct form) of men of the higher ranks. Research to this point had necessarily been limited to studying the remains of the friendless wretches dying alone in hospitals or those recovered from battlefields; what was lacking was data from the crania of individuals belonging to the cultured class—scholars, writers, industrialists, politicians, and other men of that caliber, men with healthy convolutions who in death would make positive contributions to science as they had in life to their nations.

Practical men as well as intellectuals, the members of the Society had taken measures to meet this need, supporting the formation of a Mutual Autopsy Society (*société d'autopsie mutuelle*), named "The Mutual Corpse" by one disrespectful journalist. Seventeen members had pledged their remains to Broca's Anthropology Laboratory on their decease, among them Broca's successor, the formidable Paul Topinard, who also served as director of the École d'Anthropologie and editor of the *Revue d'Anthropologie*. These were definitely men of the right sort. The young man had taken it as a compliment when one member asked him, very politely, if he would consent to have his cranium measured.

Of course, Topinard was no Broca—who was? He did not communicate the same sense of massive self-assurance and authority that had struck all who had met the Founder. An air of distrust and presumption had settled like a veil on Topinard's face. Conscious, like everyone else, that leadership had been transferred to him rather than earned, that he was not the founder but the successor, he seemed determined to overcome all objections by mentioning Broca's name at every turn, praising the work Broca had launched, promoting the ideas Broca had developed, and even growing his whiskers in an apprentice version of the way Broca had grown them. But the differences were obvious to all. Broca's magisterial gaze had been fixed on things unseen; Topinard, tilting his head in the same direction, had the stressed look of a man scanning the room for rivals or dissenters.

But give him credit, Topinard was the man for the job, talented as well as unfailingly loyal. As long as he was in charge, the Society would exemplify French preeminence in science. That preeminence would, the young man assumed, be based ultimately on the great revolutionary principles of *liberté*, *égalité*, and *fraternité* that France had contributed to the world. These principles had been inherited by the young man's own country of Haiti, which, eighty years before, had created the first black republic in the world, proclaiming French ideals even as it cast off French rule.

Haiti, the young man reflected bitterly, had so far been unable to realize those ideals in practice. Still riven by absurd conflicts between "Negroes" and "mulattoes," as though these were separate and competing species, the nation seemed doomed to repeat the same conflicts over and over in slightly different forms, while France, progressing more rapidly, was thinking of itself in more spacious and cosmopolitan terms. Just recently, the great historian Jules Michelet—a steadfast friend of the black race—had published the first volume of his history of France, declaring that the French

FIGURE 4.1. Paul Broca. Public domain. Source: Wikimedia Commons.

nation was defined not by its biology but by its character, and especially by its singular capacity for assimilation and absorption. Michelet had shown how the primitive Gaels had mixed themselves with other races to produce a nation bound together not by some primitive fantasy of the connection between people and land but by a fully conscious collective commitment to freedom and autonomy. France, he said, was composed of

> race upon race, people upon people. . . . France has formed herself out of these elements. . . . Let us not give too much importance either to the primitive element of the Celtic genius, or to the additions from without. The Celts have contributed to the result . . . so have Rome, Greece, and the Germans. But who has united, fused, converted these elements; who has transmuted, transformed, and made a single body of them; who has eliminated out of them our France? France herself, by that internal travail and mysterious production compounded of necessity and liberty, which it is the province of history to explain.[2]

FIGURE 4.2. Paul Topinard. Public domain. Source: Wikimedia Commons.

What a grand, enlightened, and emancipatory vision! Why had this precious wisdom not been left to the people when the French departed from Haiti? The French had left the nation an educational system that included Greek and Latin, but they had absconded with the purpose of learning, the commitment to liberty achieved through national unity. There was nothing in Haiti that could compare with the French nation as Michelet had described it. Why was that? What was there, he wondered, about the people of Haiti that seemed to make them incapable of realizing in their own land the noble ideas the French had bequeathed them?

The one disharmonious note in Michelet's great threnody, the young man thought, was the mention of "France herself." Why think of the nation, especially a great and powerful nation, as a woman? Nor was this a mere blunder or accident. It seemed to be a national weakness of "La France," a weakness that might in fact be feminine in origin. Even in the ranks of the Society, he had heard, there was a woman who had not merely been permitted to join but who had actually addressed the group, behaving just as if she was a man. More than presumptuous, he found this ridiculous. Was her presence tolerated as a kind of amusement, like a court jester or an organ-grinder's monkey?

The woman in question was apparently neither amusing nor womanly, and certainly displayed no feminine eagerness to please. The young man learned that she was in the habit of openly challenging the other members, her dense and disconcerting interventions sprouting parasitic branches in all directions. He was reassured to learn that his view of the matter was not his alone. Once, when she insisted on asserting her "feminism," she was censured by the great Broca himself, a rebuke that ten years later was still the only incident of its kind in the Society's history.[3] That must have been a fine and stirring occasion, a commanding assertion of clarity and authority!

There was certainly no possibility that this woman, whoever she was, could compare to the women of Haiti, especially the irresistible mulattoes, with their grace, their liveliness, the elastic freshness of their shapely bodies, and that elegant walk of which only a girl of the tropics knows the secret. Whoever she was, this woman certainly had none of that, and none even of the beauty of the typical Parisian woman. Women in Paris, the young man had noticed during the few months of his residency, were more attractive than those in the provinces, but, he also noticed, not all Parisians were attractive; indeed, some, like her, were actually ugly! So what was she doing in this community of the intellect? Her presence could only be accounted for as a demonstration of the supreme confidence of the members, their generous tolerance of all points of view even when they issued from dubious sources—or perhaps, as a small proof of the amalgamative greatness of "France herself," which could absorb, transform, and dignify anything it embraced.

But . . . the difficult thought occurred to him that his own presence might be considered an instance of French generosity toward an anomaly. He was acutely aware of the many differences between his own embattled little island nation, where disputes were still settled with guns or knives, and the hard-won but established serenity of the Third Republic. In fact, his very presence in Paris was a consequence of those differences. There had been an unfortunate sequence of events in his country that was not uncommon or successful enough to call a civil war, or even organized enough to write an account of, in which armed men had come out of the hills and set a few buildings on fire. The Capitol had been ransacked while the police looked on, a general was killed, everybody was running around and shouting. In the aftermath of what acquired the pathetic and quickly forgotten name of the "liberal's rebellion," he—a prominent "liberal"—and President Salomon decided that it would be better for all concerned if he left the country. While being in fact an exile, they agreed, he could serve in a "diplomatic" position in Paris.[4]

Nobody here in Paris, among this democracy of the elite, knew or needed to know anything of that. Here, matters of difference were approached in a far more civilized manner, as subjects for learned inquiry. Politics and personal enmity were set aside in a common quest for knowledge. The spirit of free inquiry and rational debate dominated in a community in which everyone was, in principle, equal to all others, to be judged only by the quality of one's thought or the integrity of one's research rather than the color of one's skin or the shape of one's nose.

And yet, the young man noticed, color and noses seemed to matter, sometimes greatly. Some French citizens seemed to feel that their superiority was only partially derived from their ideals, their history, and their achievements; or rather, they gave the impression that all those things were also attributable to their color, which they regarded as the visible sign of other differences that were part of the nature of things. Such differences, some were saying, entailed rights and privileges to be sure, but also obligations that a great nation was proud to meet. In acquiring as colonial possessions certain territories in Africa and Southeast Asia—so many that France had become second only to Great Britain in the extent of its empire—France had also assumed responsibility for what a speaker in the Chamber of Deputies had just described as the "*mission civilisatrice*" through which France could, by spreading the French language and Catholicism, throw a benevolent light on the dark places of the world—the places, that is, populated by dark people. The "superior races," this man had said,

> have a right because they have a duty. They have the duty to civilize the inferior races. . . . In the history of earlier centuries these duties, gentlemen, have often been misunderstood. . . . But, in our time, I maintain that European nations acquit themselves with generosity, with grandeur, and with sincerity of this superior civilizing duty.[5]

In the next sentence the speaker turned his attention to the economic and strategic importance of ports.

And now the young man was hearing talk of a conference to be held later in the year in Berlin where the European powers would divide up the entire continent of Africa, distributing responsibility for civilizing uncounted millions of people. Well, he thought, the world is what it is, and some nations undoubtedly need assistance; but at least here, among these distinguished members of the Anthropology Society, the quest for truth rather than power and wealth prevailed.

The meeting commenced.

Nothing in the political or intellectual culture of Haiti had prepared the young man for the spirit of consensual solemnity that prevailed in the room as the items on the agenda were raised, duly considered, and laid to rest. In order: a formal presentation about the recent West Africa conference; a report on ancient superstitions still surviving among the Bretons; a report on a recent conference held in memory of Darwin; a note on the evolution of the eye in lower animals; and a discussion of the uni-discoidal placenta of the mandril. All solid work, received with appreciation and entered into the record.

Spirits rose at the announcement of the winner of the Broca Prize, which was given to M. Testut in recognition of his extraordinary essay "Muscular Anomalies in Man Explained by Comparative Anatomy." A superb piece of research, undoubtedly, but the young man found himself even more interested in the two other finalists for the prize: Collingnon's "Anthropometric Differences of the Leading Races of France," and especially Chudzinski's erudite discussion of "The Anatomy of the Negro."[6]

This last was a subject he knew, or thought he knew, something about. But from the comments being made around him during the discussion, the young man began to wonder what his distinguished colleagues knew or thought they knew. The anatomy of the Negro was being described as if it was not only different from other anatomies but deeply different, with the implication that the differences did not stop with anatomy. The premise, presumed rather than argued, seemed to be that anatomical features were among the many markers of intellectual and even moral characteristics that were intrinsic to the person, and, when properly organized and generalized, to the entire race. The goal of science—of the Society—was therefore to identify as precisely as possible the physical markers that might correlate with deeper differences and go from there. To his horror, he saw that where they were going was toward a statement of the permanent inequality of the races, with the Europeans on top and the dark peoples of the earth on the bottom.

Since all these men were among the most powerful intellects in the world, they were able to generalize the question in order to raise other issues of even greater consequence than the measurement of limbs or skulls, characteristic smells, or the composition of blood. Were Negroes truly human in the same sense as whites? Might they be a different species? What was the true meaning of words such as species and race after all? If these simple and fundamental terms could be properly defined, other questions might answer themselves, but, remarkably, the terms seemed to resist any reduction to clarity despite or perhaps because of the members' best but conflicting efforts. A confusion about terms both reflected and produced confusion in thinking such that one could not even be certain which confusion one was confused by.

The temperature of the meeting rose. An impassioned but ambiguous intervention from Topinard, speaking from the chair, was met with equally impassioned rejoinders from others in which, to the young man's eyes, the object sometimes seemed to be not the inequality of the races but the unsuitability of Topinard himself to the position he held. The meeting ended with everyone upset and the question unresolved.

Humiliated by the discussion, the young man was almost embarrassed by it as well. Beyond the insult to his person implied by the enumeration of what were in effect his own deficiencies, there was the scientific ineptitude. These eminent men, recognized leaders in their field, did not know the meanings of their most basic terms, and could not even agree on their disagreement. The commanding presence of Broca, the young man sensed, was acutely felt in its absence.

For a brief, bold moment the young man considered rising to request a debate on the question of the advisability of dividing humankind into superior and inferior races. But he hesitated, deciding that he was not ready to challenge the group he had just been invited to join. Before he even thought about publicly contradicting Topinard, Joseph Deniker, Jean-Louis Armand de Quatrefages de Bréau, Gustave Le Bon, or any of them including the memory of Broca himself—an altogether disorienting prospect that made his head swim—he would have to educate himself, rapidly and dramatically. He knew nothing of the evolution of the eye, the mandril's placenta, or the folk beliefs of Bretons. His only interest was in the subject of race, a subject on which his views, if known, would be considered not controversial or even interestingly wrong but simply irrelevant, as peripheral to science as the folk beliefs of the Bretons. A direct repudiation of the scientific authorities by a nonscientist such as himself was doomed and might result in a second censure by the Society.

The young man realized that he had to learn a new and difficult language. He had to equip himself to think about human beings in a different way, not as people like or unlike oneself with whom one had certain kinds of relations, but as instances, representatives of categories, bits of evidence, sources of knowledge about species, races, nations, types, varieties. To do this, he had, like a child, to begin at the beginning. This meant Broca: he had to know Broca as well or better than his successors knew him.

Breeding at the Antipodes

The young man sought out the volumes of collected papers, *Mémoires d'Anthropologie*, that Broca had begun to publish at the end of his life.[7] These were impressive, even monumental volumes, offered to the world as a record of a life rich in intellectual drama and overmastering achievement—a life of exploration, really, in the remote provinces of human understanding.

Eagerly, he opened the first volume, five hundred and forty pages of essays, mostly on the "descriptive geometry" of the skull, a branch of mathematics that according to Broca would provide a grounding for an altogether new way of thinking about human beings, as sources of numbers that would reveal categories of nature imperceptible to the eye. The man was a genius with skulls. While impressed by the industry of the American craniologist Samuel George Morton of Philadelphia, Broca had immediately seen the limitations of the goniometer, and of any method that proceeded with a single instrument of measurement. Broca was also unimpressed by the common use of projections, as, for example, of the "auricular point," which is nothing more than the projection of the external auditory canal on the median vertical plane—clearly a dead end.[8] Even the great Blumenbach, operating in the uncertain early morning light of real science, had no proper instruments and never attempted sophisticated measurements, contenting himself with dimensions and angles visible to mere observation.

Hampered by their lack of advanced technology, the pioneers of craniology were misled into thinking that visualizations alone would yield solid information. All their results could now be replaced by data generated not only by the goniometer but also by the stereograph, kephalograph, geometric diagraph, and other instruments Broca was inventing—over thirty in all. In time these instruments would produce reliable statistics about the cephalic index, the auricular angle, the longitudinal and parietal diameters, the zygomatic and occipito-frontal arches, the minimum and maximum transverse diameters, the antero-posterior vertical curve, the transverse bi-auricular curve, the orphio-spinal angle, the anterior-posterior iniac diameter, the naso-basilic line, the sub-auricular vertical diameter, the horizontal periphery, the internal capacity, the complexity of cranial dentures, the degree of facial prognathism, and the anterior and posterior facial projections, not to mention the nasal index, which permitted a classification of skulls as leptorrhine, mesorrhine, or platyrrine. From the limited number of skulls to which he had access (the Mutual Autopsy Society not yet having been established), Broca himself was beginning to develop a statistical base on which future science could work. He ranged his collection of skulls into series—eventually sixty of them—in addition to identifying thirty-four shades of skin color. A formidable project, and yet this was just the beginning, for the domain of the measurable and categorizable was limitless.

The prose in Broca's *Mémoires* was like nothing the young man had ever read, a creolized mixture of French and Latin, a patois peculiar to the Anthropology Society. But somehow, these leisurely, even dilatory sentences humping along in their polysyllabic way managed to tell stories of extraordinary scientific heroism taking place in the quiet of the laboratory. To be sure, the story, centered as it was on skulls, existed at a certain remove from the world one actually lived in, but this remove ensured objectivity and conferred conceptual power.

And in actual fact there were real-world applications available to the determined investigator. For example: the Basques, a tiny, embattled people with a peculiar language, existing in a little pocket in southwestern France and northern Spain, claimed an ancient independence from these two nations on racial grounds, saying that they were an autochthonous pre-Celtic group. Some of the evidence for racial difference was linguistic, but it was possible to put the entire discussion on a firmer footing by acquiring a large number of Basque skulls and devising experiments.

You could, for example, construct a horizontal board graduated on two scales, placing a nail at the point of common zeros; you could then position a Basque skull on the board, with the nail extending through the occipital arch tangent to the anterior edge of the hole. Thus would be created two perpendiculars, one of which passed through the alveolar point and the other through the most posterior part of the occipital arch. The distances between the bases of these two perpendiculars and the position of the nail at the common zeros would give the aft and posterior projections, the sum of which would constitute the total projection of the head.

Of course, such a projection would merely provide a rough-and-ready estimate, not nearly as precise as those that could be produced by proper craniographic analysis. And one would have to examine vast numbers of skulls in a way that would distinguish skulls obtained from morgues, mass graves, hospitals, etc., and then compare them with a similar group of, for

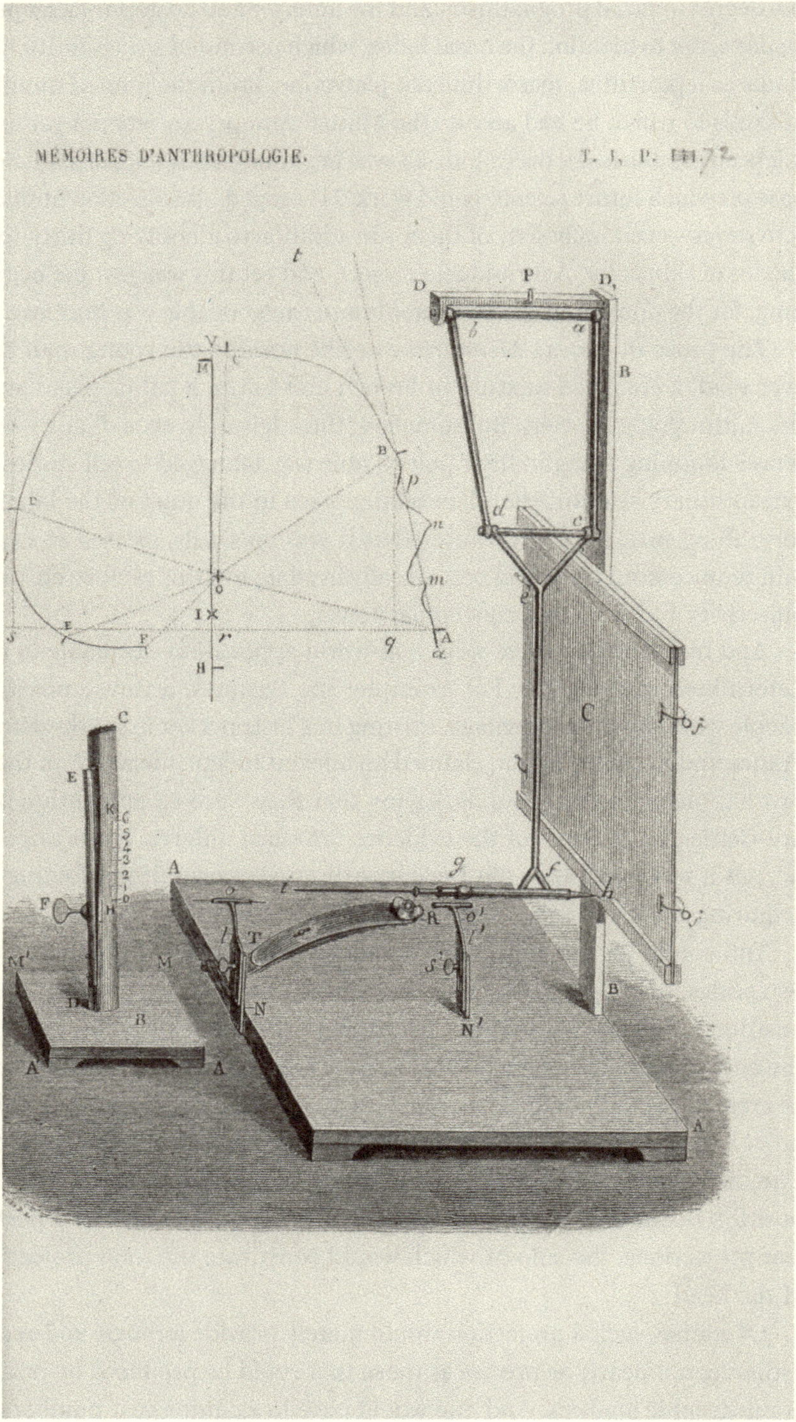

FIGURE 4.3. Craniograph, from Paul Broca, *Mémoires d'Anthropologie* (1871), 1:73. Public domain. Source: Wellcome Collection, London.

example, Parisian skulls, allowing for differences in sex, age, size, class, and other relevant considerations. But at the end of what was admittedly a very long day, you might have evidence of a kind that would help decide the question of racial distinctness and thus perhaps serve the interests of the Basque people if they turned out to be racially distinct from their neighbors in some measurable way—and if not, you might serve the interests of the French and Spanish.

The direct impact of science on the real issues of the day!

And yet it was perhaps not as direct as one might wish; in Haiti, for example, things were managed differently. Issues of conflicting interests were sorted out by other means, means that were unlikely to yield statistics other than mortality rates. The morgue would be the end of the discussion, not the beginning. The young man's attention began to wander.

He noticed Broca's deliberate use of the word race to describe human groups. As Broca explained in his introductory essay "Anthropology"—clearly intended to be a foundational account of the new discipline—some people preferred terms like variety while others preferred species. The different terms implied different theories: variety implied that all humans had common ancestors, a premise embraced by monogenists but rejected by the polygenists who were dominating current scientific discourse; while species implied that some human groups had different ancestors and different origins, a premise embraced by polygenists but rejected both by Biblical literalists and Enlightenment humanists. The use of either term would be immediately offensive to one group or the other. While Broca preferred to speak of "natural," "principal," or "fundamental" groups, he was willing to use the word race not despite but because of its uncertain meaning. It could accommodate both possibilities without prejudice, he said, "and that is why it has now prevailed."[9]

Generous and shrewd, the young man thought; but why not take a position on this most important of all questions?

It was not until the third volume of Broca's *Mémoires* that the young man found something he thought he might fully understand: a massive compilation of essays from 1858–60 titled "Memoirs on Hybridity."[10] If there was anything Haitians understand, he thought, it is hybridity. But in Haiti, reflection on the subject, such as it was, did not generally begin with "animal hybridity in general" or conclude with a consideration of "the question of the plurality of human species," as the title of one of Broca's essays indicated.[11] But this was apparently the Parisian way: the path to human self-understanding began with mules, hinnies, and other oddities and ended in abstractions and questions without answers.

In fact, Broca began his "Memoirs on Hybridity" not with animal hybridity "in general" but in curious fashion with an animal he named the leporid, a cross between the distinct species of the rabbit and the hare. It seemed that while on vacation in 1857, Broca had met a farmer who told him that he had in his possession such creatures that had procreated for seven generations. For over a century, many scientists had believed that species were unalterable and the test of species identity was fertility: if creatures could procreate, they were of the same species; if not, not. The theory had the benefit of supporting and being supported by the notion that the animal species were created by God and descended from a "unique and primordial couple," but it could not account for leporids, which, from the Biblical perspective, could only be seen as inexplicable freaks, sports, violations of nature.[12] A single generation might be a one-off, like a mule, Broca thought, but seven?

In the presence of breeding leporids, Broca had sensed that he was standing on the threshold of an entirely new understanding of nature. Rather than thinking of leporids as freaks, he asked, why not revise our understanding of the law of nature in a way that might accommodate them? Our thoughts about nature should not be clearer or more precise than nature itself. The very principles of nature were fecundity and inventiveness, and hybrids demonstrate these principles even more eloquently than conventional offspring. We must, he said, begin with the premise that any theory of species that rejects hybrids is out of step with nature. In fact, he said, we must begin with the humble recognition that we do not know what a species is. (Now, more than a generation later, the young man reflected, his successors were still in the dark but lacked humility.) Progress in understanding nature should begin with anomalies such as the leporids and build the theory around them rather than starting with some hypothetical norm or type.[13]

In a tone of worldly Parisian sophistication, Broca offered an inventory of anomalies of the sort well known to country people: dogs and sows, locked together for days at a time; hens and rabbits "and the processes, as ingenious as they are effective, which they imagined to satisfy their passion"; a scandalous hen "who prostituted herself with all the roosters in the place while her duck was out for a walk"; a "medium-sized dog taking as a mistress a Guinea goose to whom this eccentric union was not at all displeasing"; a female parakeet who, though "married to a male of his species, fell in love with a singular passion for a canary . . . and committed numerous adulteries with him."[14] Species themselves demonstrated no respect at all for the "law of species," but the fact that none of these couplings produced offspring implied a hidden structural resistance equivalent to what Morton had described in psychological terms as a "repugnance" to their unions that was lodged not in the confused brains of animals but in the nature of things. This resistance, Broca concluded, is the key fact in determining the nature of nature's categories.

Unpunished rural irregularities suggested to Broca that phrases such as "the law of nature" or "natural law" were imprecise and misleading. Nature did not have the character of a law, an imperative violated only at cost. Nature favored regularities but tolerated exceptions, singularities, and minor transgressions in the manner of a watchful but indulgent parent. What we call species and think of as sovereign and inviolate entities might be considered by nature as neighbors separated by a low fence or an ambiguous property line. The leporids, for example, seem to have obtained nature's perhaps grudging or perhaps winking assent to an almost-incestuous union that while irregular and certainly not to be condoned was not absolutely forbidden; the offspring could not be denied.

And so, Broca announced, a new understanding of nature unfurls before us. Nature expresses its preferences not by imposing iron laws or issuing unanswerable diktats but by constructing guardrails, issuing warnings, exerting pressures, neglecting to lock doors, and so forth. Nature permits and respects exploration and experimentation, even occasionally rewarding innovation with offspring. Exceptions or anomalies can range on a scale from the free fertility of the leporids through various forms of partial or diminished fertility and on down to total infertility, cases in which an exasperated nature draws the line.

This is how nature works, Broca said, and with this understanding in mind, we can now approach the subject of human variation, or what has been called race.

The Americans, Broca thought, were particularly impressive on these subjects. Of particular importance was the work of the "learned and respectable" (elsewhere the "celebrated" or "illustrious") Morton, the unquestioned pioneer in the study of skulls. Brilliant theoretical work building on painstaking empirical studies! But the project was sadly incomplete, for history had intervened, making it impossible for the American School to realize its mission. Scientific work requires philosophical calm, and what kind of progress can be expected when politics and the church get involved or when armies presume to settle questions that ought to be left to science? There was actually a moment when political debate on the destiny of Negroes in America seemed to rest on the opinion of *legislators* on the nigrifying influence of the African sun![15] No, Broca thought; science makes its greatest contribution when it is completely independent of such forces.

Reading through Broca's volumes, the young man was deeply impressed by Broca's scientific integrity and thrilled by the emancipatory implications of his reflections. From the leporid, Broca had leapt to the universal potential for hybridity in the natural world, from which point he argued for the biological dignity of the mulatto, even suggesting that all of humanity was hybrid in some degree. This changed everything. No longer the anomaly, the hybrid—the mulatto—could now be considered the new norm. This, the young man decided, was a scientific and humanitarian triumph with great implications for humanity beginning with his homeland of Haiti, a living laboratory for the study of hybridism.

In fact, in the longest essay in Broca's third volume—a mighty production clearly intended to be a decisive summative statement called *On the Phenomena of Hybridity in the Genus Homo*—Broca actually mentioned both Haiti *and France* as "hybrid nations."[16] The young man gave one passage his closest attention. The question was whether mulattoes produced by the union of a white and a Negro "of pure blood" were perfectly "eugenisic" or fertile, or whether they were biologically compromised and blocked from reproducing.[17] ("Eugenesic," the young man thought—an interesting word. Had he not heard something like it in discussions of the work of someone named Francis Galton?) The question could not easily be settled by observation, Broca said, because we have no way of observing a population of interbreeding "first degree" mulattoes. Haiti might be a candidate for research, but the island was populated by "mestizos of every shade, and even if this hybrid nation were to subsist in perfect prosperity—which it has not—during several generations, the unlimited prolifickness of mestizos of the first degree between themselves would not thereby be demonstrated."[18] Pullulating Haiti was, in short, too disorderly to study, with everyone satisfying their passion by processes as ingenious as they were effective.

So, the young man thought, this is how my nation looks to Parisians: the place where science goes to die.

But setting that aside—how are we to think of hybridity and especially the degrees of hybridity? Several theories had been proposed; Broca, of course, knew them all. Professor Serres, for example, had noted that the union of a Negro with a white woman was frequently sterile, while that of a white man with a Negro woman was perfectly fecund. Were the two races, then, in a relation of "unilateral hybridity," like goats and sheep?

Fully alive to the gravity of the implications of this hypothesis, Professor Serres gave the following explanation:

> One of the characters of the Ethiopian race consists in the length of the penis compared with that of the Caucasian race. This dimension coincides with the length of the uterine canal in the Ethiopian female, and both have their cause in the form of the pelvis in the Negro race. The result from this physical disposition is that the union of the Caucasian man with an Ethiopian woman is easy and without any inconvenience for the latter. The case is different in the union of the Ethiopian with a Caucasian woman; the woman suffers in the act [because] the neck of the uterus is pressed against the sacrum, such that the act of reproduction is not only painful, *it is frequently nonproductive.*[19]

The young man read in a state of astonishment and dismay. He was familiar with the European view that the black man's penis was more voluminous than the white man's, but he had always believed that the difference, if difference there was—he had not had the opportunity to make extensive observations—might reflect not a natural inequality but the frequency of usage, which was surely greater in the tropics than in the more buttoned-up temperate climes.

Broca also seemed to respect "Mr. Theodore Waitz, author of a scientific treatise on Anthropology," who noted that

> *the children of a European woman and a Negro are rarely vigorous.* . . . Nevertheless, we admit this conclusion with some reserve, because the avowed unions of Negroes with white women are comparatively rare, and consequently the authors who have spoken of them could only have their inferences upon a few facts. The inverse intermixture between the white man and the negress is, on the contrary, very frequent, and as prolific in the first generation as in the direct alliances between individuals of the same race.[20]

This last point was undoubtedly true, the young man thought. In the history of Haiti, and elsewhere in the Western hemisphere, there had indeed been far more cases of white men impregnating black women than the reverse, although neither Waitz nor Broca seemed interested in the reasons for the disproportion. But the robustness of the offspring had never been questioned. What knowledge did Broca have about this matter?

The young man was even more skeptical of the credentials and probity of a man by whom Broca was mightily impressed, describing him as a "distinguished observer": Dr. Josiah C. Nott of Mobile, Alabama, in the United States. Apparently the leading theorist of hybridity and certainly "one of the most eminent anthropologists of America," Dr. Nott had proposed a distinction between mixtures of "allied" or "proximate" races that could breed freely, and those involving "remote" races that would, if continued, result in the weakening and eventually the extinction of the superior race.[21] Broca seemed to think this insight remarkably useful and praised Nott for being the first to see a number of truths now commonly accepted. Specifically:

1. That *Mulattoes* are the shortest lived of any class of the human race.
2. That *Mulattoes* are intermediate in intelligence between the blacks and the whites.
3. That they are less capable of undergoing fatigue and hardships than either the blacks or whites.
4. That the *Mulatto-women* are peculiarly delicate, and subject to a variety of chronic diseases. That they are bad breeders, bad nurses, liable to abortions, and that their children generally die young.
5. That when *Mulattoes* intermarry, they are less prolific than when crossed on the parent stock.
6. That when a *Negro* man marries a white woman, the offspring partake more largely of the Negro type than when a Negro woman marries a white man.
7. That Mulattoes, like Negroes, enjoy extraordinary exemption from yellow-fever when brought to Charleston, Savannah, Mobile, or New Orleans.[22]

So, the young man reflected, according to one of the most eminent anthropologists in America, some mixed-race unions are respectable, robust, and, like leporids, can be considered part of nature's benign plan—but definitely not those between Caucasian women and Negro men, which are degenerate, unnatural, and punishable by death!

Who could possibly credit this mongrel mix of fact and fiction? And yet to the young man's horror and amazement, Broca, whose abolitionist sentiments were well known, seemed to have swallowed it whole. He even belabored the comparison between Negroes and primates, pointing out that while the zoological distance between the two is great—Thank you!—still, the "physical conformation of the Negro is in some way intermediate between the European and the monkey." For proof, Broca pointed to the Ethiopian foot, which is "somewhat reminiscent" of the same appendage of the monkey.[23]

Broca cited all these authorities—Mr. Alex Harvey, Dr. Carpenter, Count Strzelecki, Professor Goodsir, M. de Fereycinet, M. Jacquinot, M. Omalius d'Halloy, Mr. Sagot, Mr. Cunningham, M. Henricq, Mr. Lesson, and many other researchers and savants in Mauritius, Senegal, Panama, Java, Peru, Nicaragua, Tasmania, Australia, the United States and elsewhere but most particularly and most respectfully Nott—in support of the conclusion that Nott had stated with great economy and force: that while some groups were proximate to or allied with each other, the differences between the "pure" or "fundamental" races such as Saxons and Hottentots were so profound as to constitute species distinctions.[24] In these cases and only in these, nature imposed itself in the form of sterility. The infertility of such unions clinched the case that the Negro foot had merely suggested.

The question of the unity or plurality of the human species that had engaged philosophers, theologians, and scientists for nearly a century could now be laid to rest: Europeans and Negroes were so radically different from each other that they must be different species. Polygenesis was the only theory a scientist could respect.

The young man was appalled, not least by Broca's frequent assurances that he was not attempting to establish the innate superiority or inferiority of this or that race but was only inquiring as a scientist into the question of whether Ethiopians or Caucasians, for example, could have the same origin. Well, the young man thought, that is at least a legitimate question—but then what process of scientific reasoning could possibly account for the smugness and condescension of this passage?

> What are the two races forming the extremes of the human species? Several English authors express the conviction that the Anglo-Saxon, or rather the Germanic race, to which they belong, is the first race of humanity. M. Alex. Harvey is even pleased to believe that Providence has created it to rule all the rest. Patriotism is a virtue which is entitled to our esteem. We shall, therefore, not attempt diminishing the satisfaction of our allies across the straits, and we shall, at any rate, acknowledge that the race which has produced a Leibniz and a Newton is inferior to none.
>
> At the extremity of the world, and nearly at the antipodes of Great Britain, the English have been for more than half a century in contact with the Melanesian races, and specially with the Australians and Tasmanians. The relative degree of inferiority between these latter races, which differ sensibly in their physical character, may be open to discussion. It is, however, generally admitted that they are inferior at least to all other races who have come in permanent contact with Europeans.

> The Hottentot race, which has long been considered to occupy the lowest degree, is evidently superior to them. The Hottentots, though refractory to education, have, at least, shown some degree of improvability, while the Australians seem absolutely incorrigible savages.

And what, finally, was one to make of Broca's final word on the subject of hybridity, offered as proof of the scientist's ultimately philanthropic intentions?

> . . . the polygenist doctrine assigns to the inferior races of humanity a more honourable place than in the opposite doctrine. To be inferior to another man either in intelligence, vigour, or beauty, is not a humiliating condition. On the contrary, one might be ashamed to have undergone a physical or moral degradation, to have descended the scale of beings, and to have lost rank in creation.[25]

The young man read all this in a state of increasing disbelief and shame—not of himself, but of Broca and the Society of which he was now a member. And on that point, a new question arose: had he been elected as a member, a trophy, or an object of study?

The White Species Has Forever Disappeared

How to respond? Attack he must, but where, and who? The young man had no status in the Society; if he dared to speak he would be ignored or even censured like that presumptuous woman—or even tossed out. He hesitated to make a frontal assault on Broca given the esteem in which the founder was held and the immense positive contribution he had made by granting hybrids—some of them at least—full membership in the human community.

An alternative target suddenly presented itself. Right there, in the first words of *On the Phenomena of Hybridity in the Genus Homo*, a reference to "that very ingenious writer, M. A. de Gobineau," author of an immense treatise, *Essai sur l'inégalité des races humaines*, that had appeared decades ago to little fanfare. Just recently a portion of this multivolume work had been reissued, again to little fanfare, perhaps because the author had almost perversely insisted in his preface that he had altered none of his earlier views.[26] One of those views was Broca's starting point, the beliefs—mistaken, Broca would demonstrate—that hybrids were degenerate forms and that the tendency of racial mixing was irrevocably downward, with the superior race becoming corrupted. Broca, the humane

FIGURE 4.4. Arthur, Comte de Gobineau. Public domain. Source: Wikimedia Commons.

liberal, intended his work as a triumphant scientific refutation of such reactionary pessimism.

An impoverished aristocrat with a dubious claim to his title and a pronounced dislike of cities, Gobineau was, the young man understood, held in low regard by the members of the Society, who supported an aristocracy of the intellect but no other kind. But the subject of the inequalities of the human races that Gobineau had raised in a sprawling and tendentious way was in fact the focus of the Society's research. All those studies of the Negro's larynx, the recent photographs of the Neo-Caledonians, the report on a newly discovered prehistorical hydrocephalic skull, the study of Chinese fingernails—all of it was ultimately conceived as a contribution to the study of primordial racial characteristics, which would inevitably support the identification, description, and, necessarily, ranking of races, with people like the members of the Anthropology Society somehow winding up on top, and surprise, surprise, people from places like Haiti on the bottom.

It seemed to the young man that while the members of the Society may have despised Gobineau personally they were following him intellectually, discovering new ways of continuing Gobineau's obsession with race, race-mixing, and degeneration while trying to preserve some principle of pride for people like themselves. Attacking Gobineau's racial arguments directly would be a dangerous undertaking for the young man; Gobineau's more general positions might be a safer target. A thought rapidly crystallized into determination: if he could develop his own scholarly critique of Gobineau, his criticisms of Broca might be seen as friendly amendments rather than as attacks. If he could negotiate this delicate dance of agreement and refutation, he might both make his case and survive.

Thirty-three years of age when such thoughts occurred to him, Anténor Firmin—the proximate model for the young man described in the previous pages—spent the next ten months in a rare frenzy of reading and writing, producing a four-hundred-page tome with a title—*De l'égalité des races humaines*—intended to identify, evoke, and challenge the man he saw as a primary antagonist.[27] He presented it to the Society in late 1885, with the inscription "Respectful tribute [*hommage respectueux*] to the Anthropology Society of Paris"—which, however, received it in silence. Never mentioned in the official proceedings, Firmin's book was, from the perspective of the Society, impudent, misguided, altogether regrettable, and therefore negligible. From the perspective of the twenty-first century, it is an astonishing text—a comprehensive and relentlessly detailed indictment of European intellectual culture over the previous century as a monument to the principle of racial inequality and specifically of Negro inferiority produced by a young Haitian inspector of schools with a law degree and a history of ineffective political agitation but no prior record that would indicate that anything like it was in the offing.

Firmin saw Gobineau as a supremely repulsive and strangely limited figure—an aristocrat obsessed with the loss of traditional hierarchies and values, a loss he attributed entirely to a single overarching cause. The "racial question," Gobineau wrote, "overshadows all other problems in history."[28] History for Gobineau was a vast festival of degeneration, with superior races conquering their inferiors but becoming polluted in the process. All other causes of social disintegration, including revolutions, natural calamities, wars, fanaticism, luxury, corrupt manners, loss of religion, or bad government were merely secondary or downstream effects of a racial corruption that was inevitable and inexorable. Whatever the

FIGURE 4.5. Anténor Firmin, 1889. Public domain. Source: Wikimedia Commons.

proximate cause, the ultimate effect was the same: "a confusion which, like that of Babel, ends in uttere [*sic*] impotence . . . the abyss of nothingness whence no power on earth can rescue them."[29]

But why, beyond a simple organic desire to take such opportunities as proximity and genital compatibility granted, was humanity committing a long suicide? Gobineau had an answer. Inferior races with nothing to lose were particularly susceptible to the force of attraction, leading them to overcome the natural repulsion of all races for miscegenation and breed with their betters who, despite their superiority, could not be expected to resist every temptation placed in their path. All too frequently, the superior races set aside their natural disdain, mixing their blood with their inferiors, blurring and eventually eliminating the precious distinction between high and low. No meaning or value could possibly be attached to this immense process, which ground on at all times and everywhere. In local battles, the best or purest race might triumph, and civilizations might arise and flourish for a time; but in the long run, degeneration itself was the ultimate victor. The very success of the white race in conquering other races ensured its defeat in this larger struggle.

Nothing was spared Gobineau's profound pessimism. The democracy that Firmin considered the great achievement of the modern age was to Gobineau a disease with a high mortality rate for which no cure was even imaginable. As he began to take in the entirety, Firmin saw Gobineau as the voice of the unregenerate past and his book as a universal eulogy, written as if from within the tomb of culture and history. In a sense, Gobineau was quite original. Others who denied the common humanity of the world's peoples recommended nationalism or militarism as ways of preserving necessary distinctions. Gobineau did not advocate anything at all.

Firmin had never read anything conceived on such a scale; nor had he ever encountered such a one-dimensional mind. And yet, with a deepening immersion in the world of the *Essai*, he began to sense that although Gobineau seemed to be obsessed with race, he was in some ways far less dogmatic than many others. For one thing, Gobineau conceded the possibility of a common origin as described in the Bible.[30] Gobineau acknowledged, too, that all peoples think of God, Adam, and Eve as being of the same color as they are, and that science cannot disprove any such theory.[31] Moreover, Gobineau despised slavery and condemned those who engaged in it, and attributed racial differences not to nature but to climate and other factors that had worked on the progeny of Noah's sons. Overall, Firmin had to admit, there was much to praise even if the praiseworthy was immured in a gigantic slab of bigotry.

Firmin recoiled, however, from Gobineau's account of the three "original types" of humanity derived from those sons of Noah. Gobineau began at the bottom, with a series of statements that, Firmin felt, condemned the author more than the subject: that the "dark" or "Hamitic" race is even today a civilizational nullity standing "at the foot of the ladder"; that its inferiority—indeed its animality—is stamped on its pelvis and its "low receding brow"; and that it is so depraved that it is disgusted by nothing and will eat carrion. The innate repulsion felt by white people for black people was, Gobineau said, an entirely natural response given the extreme dissimilarity of the two groups and the utter squalor of the latter. Carrion-eating Negroes could never be fully civilized, and even mulattoes could only be marginally improved. Of course, a fair number of Negro chiefs might be superior to the average European peasant, but the average Caucasian was indisputably superior to the average Negro. Even the vaunted development of instinct among the savage races could only be seen as a compensatory adjustment granted by nature in recognition of the extreme limitations of their mental faculties.[32]

Wrong, entirely and disastrously wrong, Firmin felt—he himself was a resounding refutation of all of this. And yet several details seemed to strike, almost by accident, on a certain kind of truth, as when Gobineau described "an instability and capriciousness of feeling" in black people that overrides "all distinctions of good and evil." Firmin had seen countless incidents in the recent history of his country that could be characterized as capricious, and others that could serve as evidence for Gobineau's claim that the Negro is "equally careless of his own life and that of others: he kills willingly, for the sake of killing."[33] Haitian examples could even be found for Gobineau's statement that the Negro's characteristic intellectual torpor inclined him to follow leaders or ideas blindly. But, Firmin thought, is not submission to authority virtually universal? What is a king if not a master with a crown? And Haiti, like France, had mounted a successful revolution, more than one, and would you call that torpor or should it not more accurately be described as self-confidence, boldness, and faith in oneself and the future?

After what seemed to Firmin to be two needlessly vivid paragraphs detailing the characteristics of black people, Gobineau moved on to the yellow race—a group comprised of Altaic, Mongol, Finnish, and Tartar groups—in a tone that was comparably dogmatic and assured. Yellow people were the antithesis of black people and superior to them but not decisively or impressively so. The large, jutting Asian forehead did not signify stupidity, as the Negro forehead did; nor did the Asian nose or chin demonstrate the "coarse protuberances [*des saillies grossières et rudes*] that mark the negro."[34] On the other hand, however, nothing about the yellow race suggested that it could compare with whites, and many comparisons between blacks and yellows did not seem to favor either race. If black people are unstable and extravagant in their impulses, yellow people are slack and inclined to obesity; if black people are at the mercy of their desires, yellow people are peaceable to the point of apathy; if black people forget their actions as soon as they take them, yellow people are obstinate and vindictive.

In general, Gobineau concluded, the yellow race is intellectually mediocre and morally undeveloped, incapable of real achievement, given to groveling vices rather than high crimes, and above all desiring to live in the easiest and most comfortable way possible. While yellow people might be honorable functionaries in a civilization, they could never create one.

And so at last, "we come now to the white peoples," whom Gobineau depicts as at once the golden mean between unworthy extremes, the majestic peak to which the other races are foothills, and the norm from which they represent deviations.[35] All civilizations worthy of the name derive ultimately from the white race, Gobineau said, and none can exist without its help. The best elements of any civilization are those that have derived most directly from this splendid race, and a society is great only insofar as it preserves the blood of the noble ancestors who created it.

The virtues of that ancestral nobility were, for Gobineau, manifold. Possessed of great vigor, endurance, intensity of will, and superior intellect, white people have achieved great things. Utilitarian but also imaginative, whites have proven themselves capable of extraordinary inventions that black people could only imitate and yellow people only utilize. Restlessly seeking perfection in all things, the white man has always displayed an "extreme love of liberty," and is "openly hostile to the formalism under which the Chinese are glad to vegetate, as well as to the strict despotism which is the only way of governing the negro." On those occasions when they have been forced to be cruel, white people, unlike the other races and very much to their credit, have been "conscious of their cruelty." Loving life more than the others, the white man yet discovers reasons for sacrificing it, often from a sense of honor—a word unknown to the black and yellow races but one that "has played an enormous part in the ideas of the [white] race from the beginning." The difference in beauty between the Europeans and all other races is indicative of a comprehensive superiority that is "rational, logical, permanent and indestructible."[36]

Arrogance, Firmin thought, does not begin to describe such delusional thinking. How could an intelligent and learned man entertain such thoughts, unless . . . unless there were other perhaps more defensible thoughts, not expressed directly, behind and beneath them? Firmin made a conscious effort to set aside the racial insult in order to focus more closely on this notion of honor that seemed so important to Gobineau.

In the stratified world of Gobineau's imagination, a sense of honor was the privilege of aristocrats, and the white race was the aristocratic race. Aristocracy and fair skin seemed to be, in Gobineau, variants of the same concept so that each nation's aristocracy represented the "whites" of that nation, the genetic repository of the nation's distinctive values, principles, and traditions, the embodiment of the nation's concept of honor. Perhaps, then, it was not only the loss of lineal integrity that accounted for the sense of hopelessness that hung like a shroud over the *Essai*, but the loss of honor, the foundation of every stable society and of civilization itself.

This suspicion was supported by the fact that France, not Haiti or any other country, was cited as a particularly appalling example of the loss of honor that attended racial dilution or corruption. The Teutonic or Frankish aristocracy represented by the Bourbon dynasty had, Gobineau argued, been displaced as a consequence of the Revolution by the common people, the inferior Celts and Gauls that the Franks had conquered in the fifth century. The vital, invigorating, and organizing principle of inequality that defined the nation had been destroyed at a stroke in a misguided enthusiasm for the phantom ideals of freedom, progress, and liberty. The destruction of the aristocracy as a ruling force both symbolized and completed the loss of the connection between nature and culture and ensured an accelerating decline.

Almost every country in the world had managed to earn Gobineau's contempt. France was just one instance of a phenomenon so comprehensive that only the Jews and Arabs provided compelling counterexamples. At several points, Gobineau praised the Jews in particular for their ability to maintain their racial purity for as long as they had, depicting them as a superbly forged race comparable to the ancient Greeks. "And what," he asked, "did the Jews become in this miserable corner of the earth? They became a people that succeeded in everything it undertook, a free, strong, and intelligent people, and one which, before it lost, sword in hand, the name of an independent nation, had given as many learned men to the world as it had merchants."[37] An impressive achievement—but confined to the distant past, in a wretched nowhere crawling with infidels.

As Firmin absorbed the difficult, despairing, self-complicating arguments of Gobineau's *Essai*, he began to see in the author a figure of great dimensions and tragic force, a man possessed of intelligence, energy, and learning who had committed himself to an idea that deprived him of all appreciation for the present and all hope for the future. The sheer magnitude of his project made him a worthy adversary, someone to be reckoned with, confronted and defeated. In his way, Gobineau was also someone to be respected, even someone to be grateful for because he seemed to anticipate and even solicit his own refutation.

Firmin had to acknowledge his adversary's virtues. Unlike anyone in the Anthropology Society, Gobineau conceded that Europeans in particular tend to be intolerant of any group other than their own, and had proven to be incapable of improving the morals of their inferiors. "Conscious of its own power and greatness," Gobineau had written, "European civilization finds itself confronted either by different civilizations or by peoples in a state of barbarism. It treats both kinds with equal contempt; and as it sees obstacles to its own progress in everything that is different from itself, it is apt to demand a complete change in its subjects' point of view. . . . [And yet] we cannot find the least trace of any moral influence exerted by her, even in the case of the peoples she has most completely conquered."[38]

Such candor in a French aristocrat compelled admiration from one who had seen so little of it in the aristocratic intellectuals of France.

Indeed, in reading Gobineau, Firmin may have experienced not a secret repulsion but a secret attraction. He may even have discerned, beneath the layers and layers of insult and wrongheadedness the distorted lineaments of a mirror image of himself as he was formulating his plans for a major intervention into the intellectual disputes at the Anthropology Society.

Like Gobineau, who had undertaken a career in diplomacy after publishing his book, Firmin was a worldly young man entrusted by his nation with diplomatic responsibilities in foreign countries. Like Gobineau, Firmin aspired to comprehensive knowledge acquired through experience and vast reading. Like Gobineau, Firmin entertained unfashionable convictions. Like Gobineau, Firmin wrote rapidly in a spirit of dogmatic assertion, driven more by his antipathies than by his affinities or enthusiasms. Like Gobineau, Firmin thought that no matter how assertive and self-congratulating their political character, the French were blind and cowardly in science. And like Gobineau, but conspicuously unlike the members of the Anthropology Society, Firmin was altogether unconvinced that the path to human self-understanding ran through the skull.

Perhaps the most surprising point of contact, to Firmin, was Gobineau's interest in Haiti, the former French possession that had proudly adopted the principles of the Revolution as its own, throwing the French out in the name of French liberty. Nobody in the Society other than Broca devoted any thought to Haiti, but if they had, they would in all likelihood not have stressed the similarity between the political cultures of the two nations as Gobineau did. In Haiti, Gobineau wrote,

> We find a society of which the institutions are not only parallel to our own, but are derived from the latest pronouncements of our political wisdom. All that the most enlightened liberalism has proclaimed for the last sixty years in the deliberative assemblies of Europe, all that has been written by the most enthusiastic champions of man's dignity and independence, all the declarations of rights and principles—these have all found their echo on the banks of the Artibonite. Nothing African has remained in the statute law. All memories of the land of Ham have been officially expunged from men's minds. The State language has never shown a trace of African influence. The institutions . . . are completely European.[39]

Independence, dignity, rights, principles—these were, Firmin felt, indeed the fundamental principles of his native land. How gratifying to see a

French citizen—an aristocrat!—acknowledging the intimate relation between the two proud nations.

Alas, Gobineau then continued his reflections on Haiti, dwelling in corrosive detail on imperfections and shortcomings. As Firmin read on, he realized that his nation was being used not as an instance of the capacity of black people to follow enlightened principles of governance but as an example of a nation laboring under a political system alien to its native genius. "Let us consider," Gobineau had written, "how [the institutions] harmonize with the manners of the people."

> We are in a different world at once. The manners are as depraved, brutal, and savage as in Dahomey or among the Fellatahs. . . . The question of cleanliness never enters anyone's head. If you wish to approach a high official in this country, you find yourself being introduced to a gigantic negro lying on his back, on a wooden bench. His head is enveloped in a torn and dirty handkerchief, surmounted by a cocked hat, all over gold lace. . . . Do you wish to question him, to penetrate his mind, and learn the nature of the ideas he is revolving there? You will find him as uncultured as a savage, and his bestial self-satisfaction is only equalled by his profound and incurable laziness. If he deigns to open his mouth, he will roll you out all the commonplaces which the newspapers have been inflicting on us for the last half-century. The barbarian knows them all by heart. He has other interests, of course, and very different interests; but no other ideas. He speaks like Baron Holbach, argues like Monsieur de Grimm, and has ultimately no serious preoccupation except chewing tobacco, drinking alcohol, disembowelling his enemies, and conciliating his sorcerers. The rest of the time he sleeps.[40]

Harsh, ungenerous, and insulting—but, Firmin might have conceded, not altogether unfounded. Civil wars, with their abrupt transitions, often produced unqualified, untrained, and uneducated officials, and neither the culture nor the climate encouraged industry, conscientiousness, diligence, idealism, discipline, commitment to abstract principles, or extended periods of earnest study. But how common were these virtues anywhere?

Firmin noticed, finally, that Gobineau was at his most penetrating on the subjects that had dominated the political life of Haiti for decades: race and racial hostility. Gobineau's contempt for Negroes as animalistic, murderous, incapable of civilization, greedy, and lazy constituted a steady ground-note in his long text; and the Negroes of Haiti were cited as irrefutable evidence of the character of Negroes in general and thus for the inequality of the human races. But to Firmin's eye the real source of Gobineau's interest in Haiti actually seemed to be the predominance of mulattoes in the life of the nation, which Gobineau treated as a local instance of the fate of all nations in which racial promiscuity has proceeded to the point where any sense of a natural racial aristocracy has been eliminated, leaving mongrels and Negroes to battle it out. "The State," Gobineau had written,

> is divided among two factions. These are separated from each other by a certain incompatibility, not of political theory, but of skin. The mulattoes are on one side, the negroes on the other. The former have certainly more intelligence and are more open to ideas. As I have already remarked in the case of San Domingo, the European blood has modified the African character. If these men were set in the midst of a large white population, and so had good models constantly before their eyes, they might become quite useful citizens. Unfortunately the negroes are for the time being superior in strength and numbers. Although their racial memory of Africa has its origin, in many cases, as far back as their grandfathers, they are still completely under the sway of African ideals. Their greatest pleasure is idleness; their most cogent argument is murder. The most intense hatred has always existed between the two parties in the island. The history of Hayti, of democratic Hayti, is merely a long series of massacres; massacres of mulattoes by negroes, or of negroes by mulattoes, according as the one or the other held the reins of power.[41]

True in a sense, perhaps, Firmin reflected, but also terribly untrue. Gobineau failed to appreciate something that was obvious to all Haitians, that the two factions may sometimes be at each other's throats but were not always so and were not even altogether distinct. Firmin himself had recently married a mulatto, who had borne him two mulatto children; he had even been labeled a mulatto by his enemies during the recent civil war. In Haiti, it was really all a matter of *plus ou moins mulâtre*.[42]

In fact, Firmin decided, the flourishing of Haitian mulattoes disproved both the argument Broca had been making that mixed-race unions had diminished fertility, and Gobineau's insistence that race-mixing represented social collapse. A Haitian in the halls of the Anthropology Society, Firmin himself might be seen as a kind of cultural mulatto, his inclusion signifying a degree at least of intellectual confidence by an organization that willingly exposed itself to transformation, incorporating the new and welcoming a future different from its past. Gobineau was completely wrong on a fundamental—really, *the* fundamental—point.

Or at least he was if he was really saying what he seemed to be saying. But sometimes, despite Gobineau's boldly declarative style, it was difficult for Firmin to puzzle out the point. Why, for example, include an entire chapter on the inequality of *languages*? And why focus, in that chapter, on the language of the "Aryans," a group so remote in time that its very existence could only be inferred by exceptionally learned scholars from scant linguistic traces deposited in Latin, Greek, Sanskrit, Persian, Celtic, and Germanic?[43] If your subject is the permanent inequality of races, why waste your breath on a no-longer-existing language spoken by a no-longer-existing race?

The Aryans were admittedly intriguing, an all-conquering horde that, in one possible narrative, had migrated from a homeland in central Asia to northwestern India, and then to Persia and the Middle East before finally settling in Europe, imprinting their language and by this means their mind everywhere they went. Their very existence was, apart from the evidence gleaned from languages and supplemented by an account in the Rig Veda (second millennium BCE) of an invasion by "Aryas" five thousand years prior to that, entirely a matter of speculation; and yet, Firmin realized, the Aryans were an absolutely essential element in Gobineau's basic understanding of the world. They had to exist, if only for the sake of his argument.

In Gobineau's mind, the perhaps-nonexistent Aryans were not defined by their language or religion but by their race. The aboriginal form of the white race, a pristine group with no admixture or impurity, the "single starting-point," "single germ," and first family of mankind, the Aryans possessed a virtual monopoly on beauty, intelligence, and strength. Everything good in the world could trace its origins to the Aryans, even history itself—he had titled one chapter "History Only Exists in White Nations." White conquests had resulted in the creation of all ten of the true civilizations in the world including the *race germanique* from which the Frankish people had descended. "All civilizations derive from the white race," Gobineau said; but over the course of time the Aryans, following a civilizational imperative to form favorable mixtures, had become so inmixed with their inferiors that their race no longer existed.[44] In essence, the Aryans had a three-part mission: conquer, breed, and die.

The logic and sequence were clear to Firmin; nevertheless, it was almost shocking to read in Gobineau's "General Conclusion" a direct statement to this effect.

> The white species, taken in the abstract, has forever disappeared from the face of the earth. After the age of the gods, when it was absolutely

> pure; after the age of the heroes, when mixing was infrequent and moderate; after the age of the nobles, when still great faculties were no longer renewed at dried out sources; it went on, more or less quickly depending on the place, irreversibly to confuse all its principles as a consequence of all its heterogeneous unions.[45]

Just as astonishing as the concession that there were no more true whites was the candid and despairing acknowledgment that all civilizations were at this late date degenerate composite forms. Every racial group was a product of aeons of what in Haiti was called *métissage*; they were all mongrels whose only connection to their origins was through their lineal aristocracy, if they had one. Gobineau himself could stand as a representative of this immense process; his claim to "comte" came from a deceased uncle and was unaccompanied by any land or fortune. Everyone—Caucasian aristocrats included—was *plus ou moins mulâtre*![46]

At this point, the only thing that could shock Firmin more deeply than the concession of universal degeneration would have been an acknowledgment by Gobineau that racial mixture could produce positive effects, that degeneration did not always lead downward. But—there it was. On the bottom of the page on which Gobineau declared that the beauty of the white race is "rational, logical, permanent, and indestructible," Firmin found a footnote: "It may be remarked that the happiest blend, from the point of view of beauty, is that made by the marriage of white and black. We need only put the striking charm of many mulatto, Creole, and quadroon women by the side of such mixtures of yellow and white as the Russians and Hungarians."[47]

True! Firmin thought. But more—far more—important even than beauty are other by-products of degeneration listed by Gobineau. Artistic genius, for example, was "equally foreign to each of the three great types [and] arose only after the intermarriage of white and black."[48] White people may be masters of form and order, but as Gobineau acknowledged, "the black element is certainly indispensable for developing artistic genius in a race, for we have seen what outbursts of fire, flames, sparks, vivacity and spontaneity are intrinsic to its soul."[49]

All the world's great civilizations bore witness, in Gobineau's account, to the benefits of mongrelization. Even the Hellenic Greeks were composites, with elements of Slavo-Celtic, Thracian, Phoenician, Arab, Hebrew, Philistine, Libyan, and Cretan. Intermarriage had produced in each of the world's civilizations results that no group could have produced on its own. In fact, the real genius of the white race resided not in itself but in its self-immolating ability to create civilizations through advantageous mixtures with other, inferior races that, left to their own devices, would be incapable of advance. “I do not deny,” Gobineau wrote, “that these are good results . . . wonders for which we must needs be thankful”—despite the ultimately tragic enervation of the great race that had been required to achieve them.[50]

It was, to Firmin, an unexpected boon to see in this most self-deluded and reactionary of thinkers such agreement with his own position on the cultural value—and even the beauty—of the mulatto. But then, Firmin wondered—if Gobineau saw degeneration as a universal disaster, and the white race as the only truly creative race, *but also* understood that civilization itself was the consequence of degeneration and that the white race could not create civilization on its own, then what did his argument amount to? What was he actually deploring and denouncing at such remarkable length?

Perhaps his target was not racial inequality at all; perhaps his argument, considered at the deepest level, did not really—or even—concern race but rather inequality *as such*. His model for an ideal society seemed to be the Hindu caste system in which everyone understood their places. Perhaps he valued aristocracy not as a source of wisdom and tradition but just for itself, as a form of authority and privilege that had some remote ancestral claim to racial purity—in the case of Europeans, to the vanished Aryan race. Perhaps this was why Gobineau was so inconsistent in his use of the word race, using it interchangeably with family, variety, group, or civilization as if he did not care about or even note the differences between these terms.

Perhaps, in other words, Gobineau's target was not lesser races, which cannot be blamed for their inferiority and which are in any event necessary for civilization. Perhaps it was not even racial mixing, which is inevitable and beneficial. Perhaps it was modernity and especially democracy, with its promiscuity, its hostility to distinctions of rank, its crippling uncertainty about who is in charge or what is worth what. Perhaps Gobineau, like Marx, was really writing about class consciousness, with the difference that Marx sought to unify humanity and so opposed principles of division like race and class, whereas Gobineau sought to divide and so favored such principles. For Gobineau, the ambiguous concept of race enabled him to suggest that hierarchy was a fact of nature and therefore deserving of respect rather than a consequence of power deserving—in our degenerate democratic and capitalistic era—only contempt.

Second to No Other Race

Absorbing huge quantities of information daily, Firmin developed his project rapidly as both a history and a general indictment of the scientific field of anthropology, which had betrayed its true promise and potential by investing itself in the subject of race. He began his book by tracing the origins of the field in travelers' narratives that included often fantastical accounts of the peoples of the world. He then showed how an era of "ethnography" based on description had been followed by an era of "ethnology" dominated by rationality and empiricism. Ethnographers looked at the general characteristics of groups; ethnologists examined the details. Ethnographers traveled the world; ethnologists measured specimens and created categories, trying to discover the correlations between, for example, variations in skull shapes and aptitudes or capacities. The culminating moment was the emergence of the anthropologist, who sought to place man in relation to other animals. Most contemporary anthropologists treated humans as animals that differed from others only by a few degrees of superiority. And this, Firmin noted, was considered the triumph of the scientific approach, a point of maturation marked for many by the emergence of professional organizations like the Anthropology Society.[51]

In fact, Firmin argued, the professionalization of science had been a fatal wrong turn. Focusing on taxonomy and considering only the physical, anthropology had blinded itself to its true mission, which was to consider man in the abstract as a singular creature both inside and outside the natural order. A true anthropology would study not just the physical dimensions of humans but also the intellectual and moral capacities that set humans apart from the rest of creation. This was not the course that had been chosen. Instead of posing real questions, such as whether all humans were capable of rising to the same intellectual and moral levels, or whether some inherited biological factors favored some and inhibited others, the anthropologist had wasted his time measuring skulls in order to determine whether or to what degree the Negro was intermediate between Caucasians and apes.

In the wake of Broca's fabulous run of technical inventions, the Anthropology Society had been swept up in craniomania. Under the banner of empiricism, scientists had abandoned people in favor of bones, veering off in the direction of trigonometric calculations applied to crania. The sapiens part of *Homo* had been set to one side, as had life itself, and human uniqueness ignored. This, Firmin believed, was a tragic mistake, for "the only message a skull ever conveys is equality in death."[52] A man is not just a bag of bones!

Not only that—*the numbers did not add up*. Unlike most other readers confronting a densely printed page of statistics, Firmin actually studied the figures and saw that the measurements of supposedly distinct racial groups often overlapped or even, awkwardly, proved the opposite of what was being claimed. The figures from the "nasal index" that supposedly constituted a racial characteristic actually suggested that the *white* race was "an intermediate type between ape and man," and with respect to the aveolo-condylian angle, Ethiopians differed very little from Caucasians.[53] Such sloppiness was symptomatic of a disorder affecting the entire project. Craniology was a surface lesion that indicated a ramifying cancer spreading unseen and undetected through all the Society's thinking. Dominated by skullwork, Parisian anthropology had become a thanatology that, Firmin strongly implied, deserved to die.[54]

The more Firmin thought about it, the more clearly he saw death at the core of the discipline. The idea of distinct and unalterable categories of humanity was a dead concept that had, in the colonies, become implicated in death, even in mass slaughter. On the premise that Africans could never manage to achieve the most basic level of civilization, Europeans had enslaved or murdered vast numbers of them, and had done so with implicit approval from the leading scientists in the world. White reason divided humanity into races not just to conquer but to exterminate. The sociologist Herbert Spencer, Firmin noted, had argued, apparently with a straight face, that

> the Hebrews, who believed they were entitled to the lands God had promised them, felt authorized, in some cases, to exterminate the original inhabitants. We too, bowing to "the manifest intention of Providence," dispossess the inferior races every time we need their territories. But we, at least, massacre only those we feel we have to massacre, and we let live those who submit to us.[55]

The "survival of the fittest"—Spencer's great contribution to the intellectual culture of the day and the source of his considerable fame as a boldly original thinker—required the slaughter or submission of the "unfit."[56]

The more Firmin learned about the most distinguished and admired European thinkers, the more he became persuaded that for many of them, vanity, arrogance, cruelty, and greed came first, and racial science, with its divisions, categories, and statistics, followed obediently after, a simpering tool of the death-machine that loudly proclaimed its neutrality, its independence, its devotion to fact, its modernity. Gobineau's premise of "an innate, original, profound, and permanent inequality among the races" was a clear and unembarrassed statement of European conventional wisdom; only slightly less conventional was the conviction that the enslaved or enslaveable were a different kind of animal from the enslavers.

Anthropology had become dedicated to this latter proposition, sometimes voiced quietly so as not to draw attention to itself. One *soi-disant* anthropologist, a former police officer named Alphonse Bertillon who had developed a system of measurements that would help identify criminals, defined the discipline as "a pure and concrete science whose goal is the complete knowledge of the human group considered, firstly, in each of its four typical divisions (variety, race, species, if called for) in comparison with one another." What, Firmin demanded, could possibly "call for" a consideration of species differences within the human? And this *policeman* was actually taken seriously—and quoted by Topinard![57]

The entire concept of race, Firmin decided, was a fraud perpetrated on a gullible public by an equally gullible and complicit scientific community, many of them in the Anthropology Societies of Paris, New York, Berlin, and London. Race was not a fact studied by science but an artifact dreamt up by science, an "absolutely secondary phenomenon" resulting from climate and long isolation.[58] In fact, there were only three basic groups, the white, the yellow, and the black. Each had adapted itself to the situation in which it found itself and could adapt to other circumstances in short order if necessity demanded. The very idea of race as a biological category ambiguously equivalent to species was an insult not just to the lower-ranked races but to humanity as a whole.

Declaring these truths, Firmin named names and spared no one. Morton, Ernest Renan, Broca, Quatrefages, Nott, Gobineau, and others—"the whole proud and arrogant phalanx of those who proclaim that the Black man is destined to serve as a steppingstone for the White man in his quest for power"—all of them stood, in Firmin's eyes, condemned as frauds and liars whose sham practices degraded the very concept of science.[59]

One member of the Society was particularly culpable—Clémence Royer, the French translator of Darwin, whose membership in the Society Firmin found deeply disturbing. She had apparently been admitted on the strength of her connection to advanced thinking, as evidenced by her long and tendentious preface to Darwin's *On the Origin of Species*, in which she had declared that "Nothing is more obvious than the inequalities that exist among the different human races." Such pronouncements were entirely in the mainstream of Society thinking; only a bit less so was her insistence that the "law of progress" dictated that the superior races would not intermix themselves with their inferiors but would "supplant" them.[60] Worse still, from Firmin's perspective, Royer had weighed in with "feminist" deliverances that were considered irritating distractions, a judgment that found Firmin and his fellow Society members for once in complete agreement.

But Firmin had issues of his own with her, for Mlle. Royer had ventured well outside the limited field of her questionable expertise onto terrain he himself commanded. On the subject of *métissage*, for example, she had once characterized the great novelist Alexander Dumas as an instance of biological confusion resulting from the mixing of a superior with an inferior race—"a maladjusted individual, or at least an abnormal one," she had written, adding that "he was a very gifted White Negro, but morally, he was a Negro."[61] Firmin registered the affront but also took the opportunity to note a significant fact: Royer acknowledged that many men of genius were of mixed race and had even suggested that their genius was produced by the mixture. "At this rate," Firmin wrote with unconcealed satisfaction, "on the basis of their psychological and moral profile alone, all celebrities will be classified Negroes."[62]

The embarrassment to the Society represented by Royer's inclusion was the inevitable consequence of a problem to which Firmin felt his French colleagues were curiously blind. She should never have been admitted in the first place. She was a woman, and for problems of any seriousness or complexity, you want men, "for only men, because of their education and their temperament as males," can see hard problems from every angle. It is "well known that women have a tendency to embrace current ideas and to perpetuate accepted notions [and] I cannot help but point out that she remains a woman." As such, she was not only bigoted but in her unworldly womanish way, blithely ignorant of the likely consequences of her beliefs, especially those concerning race. On these Firmin felt he could enlighten her. "She may indeed forget if she wishes to," he wrote with virile emphasis, "that when political and civic equality, equality before the law, is not spontaneously granted, there are Negroes who will simply use force to take it."[63]

What Royer and the others missed, Firmin believed, was the massive historical fact of black superiority, a fact fully in evidence as far back as ancient Egypt, when black people dazzled the world with colossal achievements in sculpture, architecture, and mathematics at a time when Europeans were still living in huts. A thousand years before Moses,

FIGURE 4.6. Clémence Royer, 1865. Public domain: Source: Wikimedia Commons

Egyptians were developing philosophy, science, and the arts, and enjoying "the delicacies of a refined social life." "No sculptural tradition, no school of architecture," he wrote, "will ever match the boldness of the ancient Egyptian canon with its inimitable gigantic proportions and pure lines." And Morton and Gliddon notwithstanding, the Egyptians were definitely "not a White race," for numerous depictions show a people "closer to the Black type than to the White type"—a fact that delivers a fatal blow to the myth of Negro inferiority. All the Egyptians had lacked was "the moral evolution of the Western peoples."[64]

Arguments for the contributions of black people, distributed throughout *The Equality of the Human Races*, are compressed into a single defiant statement in chapter 17, "The Role of the Black Race in the History of Civilization," which not only rehearses the contributions of the Egyptians and Ethiopians but includes several pages on the heroic example of the Haitian slave revolt and the invaluable assistance offered by Haiti to Simón Bolívar. In the past, Firmin wrote, Egypt had towered over the rest of the world; in the present, black leadership has passed to Haiti.

> Beside this example [assistance to Bolívar], which is one [of] the most beautiful actions for which the Black Republic deserves the whole world's esteem and admiration, we can say that the declaration of independence of Haiti has positively influenced the fate of the entire Ethiopian race living outside Africa. At the same time, Haiti's independence has affected the economic system and moral order of all the European powers that owned colonies. . . . The actions of the Blacks of Haiti indeed offered the most complete refutation of the theory according to which the Negro was a being incapable of grand and noble actions, incapable especially of standing up to White men.[65]

If Firmin was at all troubled by the fact that Haiti could at that moment hardly stand up to anyone as a model for a stable multiracial democracy committed to "law, justice, and freedom," he did not say so. Nor did he concern himself with the fact that ancient Egypt was a society in which most living Egyptians were compelled to toil on behalf of a few dead ones. Nor was he disturbed by the likely means by which darker Egyptian rulers acquired lighter-skinned foreign wives: he cited as a "well-known fact" that "the marriage institution was a sort of servitude of women," adding that this was actually "a natural and simple practice." In fact, he pointed out, the light skin of some wives depicted in paintings and statuary confirmed the exalted status of blacks in Egypt since "usually it is the more powerful and the more civilized people who would seize and keep the women of a weaker people."[66]

Arguing for the historical brilliance of black people, Firmin managed to defend slavery and the oppression of women. Asserting that "all men are brothers," he excluded women from the brotherhood even as he reintroduced an essentialized understanding of races as distinct categories of being. And by way of refuting the biology of the day that elevated one race over others, he proposed not equality but a different biology that yielded a different racial hierarchy. In a nine-page chapter devoted to "Aptitudes and Organic Qualities," he asserted that:

— No other race possesses a livelier intelligence, a greater faculty of assimilation, and a greater knack for progress than the Black race.

— Even if we considered only the field of science to compare the aptitudes of the different races, we would find that the Black race takes second place to no other race.
— The better and faster the blood circulates in the encephalic organ, the greater the brain's energy is. . . . It so happens that the brain of the Black man is much richer in blood vessels than that of the European.
— In the brain of the Black man the cortical substance, where most of the organ's activity takes place, is much darker than in the brain of the White man.
— What is certain . . . is that blood circulation in the Ethiopian's brain is incomparably intense.
— Everything suggests that, as a result of a more intense blood circulation, the Black man's brain has a higher temperature [than that of Whites]. . . . An intense blood circulation and a high temperature are two clear signs of great brain activity.
— We must simply conclude that the Ethiopian is endowed with the most marvelous tool to conduct any mental operation.[67]

From the perspective of the twenty-first century, Firmin's understanding looks inconsistent with itself. How could an anti-colonial intellectual prodigy and hero of modernity in so many ways defend the ideas of biological races and race ranking, see the taking of slaves as a sign of racial superiority, and practice as well as advocate misogyny in the belief that it was not only advantageous but entirely natural?

But the same challenge could be made to others whose thinking seems in the tribunal of posterity to be out of phase with itself. How is it that Broca, the first to accord full respect to the category of the racial hybrid, was also committed to what amounted to species differences among races, racial hierarchy, and Negro inferiority; or that Royer, a feminist and one of the leading experts on Darwin, could take racial inequality as an indisputable fact of nature; or that Gobineau, the "father of modern racism" as one study calls him, could also argue, to the horror of Nott and the delight of Firmin, that civilizations owed their essential and distinctive character to racial mixing? How could Firmin find greater common ground with Gobineau than with Broca? The thinking of each of these brilliant, learned, sophisticated people seems, from the perspective of the twenty-first century, to have a profoundly hybrid character.

The question for us is similar to the one these thinkers confronted: Can we respect the hybrid? Does the blending of what appears to us to be truth and falsehood, virtue and mendacity, or blindness and insight in a given understanding of race represent a correctable mistake, or is it something deeper? It is worth noting in this context that all those whose views have been taken up in this book have been confident in their possession of the truth, but no view has swept the field; all have been contested, often in the most vehement terms. We have so far failed to identify an account of human variation under the heading of race that could stand the test of time and seem equally true to one and all; all, it seems, are *plus ou moins mulâtre*.

If the facts in the case were not in dispute, the differences among the scientists and intellectuals discussed in this book might not have occasioned such extreme or complex emotions. But the facts of race have never been clear—indeed, what we have been tracking is a century-long effort to determine what would constitute a fact. Disagreement arises not from different interpretations of the evidence but from other sources altogether, from differences in the philosophical, social, historical, moral, or economic ends that the concept is meant in each case to serve. Such ends, involving one's most fundamental understandings of the conditions, limitations, and possibilities of life, cannot always be directly expressed or even acknowledged. They can, however, sometimes come to countenance in displaced forms, as in, for example, the discourse on race.

Afterlives

After he submitted his book to a decidedly unappreciative Anthropology Society in 1885, Firmin remained in Paris for several years before returning to serve in several positions in the Cabinet of Haiti. As we saw at the end of the previous chapter, Firmin was serving as Minister of Foreign Affairs in 1889 when he met the elderly Douglass, who had been sent by the American government to negotiate the acquisition of a deepwater port on the western tip of the peninsula, where the Americans hoped to build a naval base. Firmin admired Douglass deeply but distrusted the American government. Over the course of Douglass's ten-month stay in Haiti, Firmin was able to persuade Douglass to sympathize with the Haitians' hostility to the port. Opposed by the Haitians and by its own representative, the American effort failed.

For many years, Firmin worked to plan a Caribbean Confederation that would unify Haiti, Cuba, the Dominican Republic, Jamaica, and Puerto Rico. Along with Henry Sylvester Williams and Bénito Sylvain, Firmin planned the first Pan-African Congress, which was held in London in

1900. A young American scholar, W.E.B. Du Bois, was placed in charge of drafting the general report. Over the next twenty-five years Du Bois became the leading figure in the Pan-African movement.

In 1902, Firmin was a candidate for the presidency of Haiti in a campaign that became a virtual civil war. After a series of disastrously misjudged events, the "Firminist Insurgency" was defeated and he was forced into exile on the island of St. Thomas. A second and even more quixotic insurrection in 1911 also failed. Firmin had arrived by boat expecting to become the leader of his nation but was prevented from coming ashore and forced to return to St. Thomas, where he died shortly thereafter.

No biography was published until 1964, when the Haitian diplomat and intellectual Jean Price-Mars published *Joseph Anténor Firmin.* Ignored in France when it was published, *De l'égalité des races humaines* fell out of print and was almost unknown until 2000, when an English translation by Asselin Charles appeared.[68] Near the end of this book, Firmin had predicted that one day the United States, "the most progressive country on earth," might be led by a black president.[69]

Paul Broca's was one of seventy-two names inscribed on the Eiffel Tower when it was erected in 1889 in commemoration of the centennial of the French Revolution. As Firmin predicted, Broca is known today almost entirely for his work in identifying the functions of various parts of the brain, and not for his contributions to anthropometry or craniometry. And as Firmin recognized, Broca did not always insist on the premise that races could be ranked according to their cranial capacities. In an 1875 note on sexual selection, Broca echoed a statement made by Morton, allowing that as society progresses over time, brains actually become larger, adding that "these changes are the effect of education," a result Broca thought held particular promise for women.[70] Still, a century after his death, an exhibition at the Faculty of Medicine at the Sorbonne included, in addition to praise for his scientific achievements, a denunciation of Broca for having affiliated himself and the Society with "*les auteurs esclavagistes américains* (Nott et Gliddon)" (the American slavery authors Nott and Gliddon).

At the time *The Inequality of the Human Races* was published (1853–55), Gobineau was known primarily as a poet and a novelist. In addition, he had begun a career in public service in 1849, when his friend Alexis de Tocqueville, the new Minister of Foreign Affairs, asked him to become his official secretary. He became First Secretary of the French Legation in Berne, where he wrote most of his massive work. In subsequent years, Gobineau served France in various capacities, with diplomatic postings in Switzerland, Germany, Newfoundland, Persia, Greece, Brazil, and Sweden.

Gobineau was disgusted by the appropriation of his work by anti-abolitionists in the United States; if he had lived long enough, he would undoubtedly have had similar feelings about the enthusiasm for his work of German antisemites beginning with Richard Wagner and his son-in-law Houston Stewart Chamberlain. His fame came on terms he would have rejected. Long ignored, Gobineau became a thinker of recognized importance only at the end of the nineteenth century in large part due to the promotional efforts of Georges Vacher de Lapouge, a French advocate of eugenics, socialism, Aryanism, and racial ranking who became an important influence on twentieth-century Nazi racial science.

Gobineau was deeply ashamed of his mother, who abandoned her husband when Arthur was still in his teens, lived openly with her lover (Arthur's tutor), and entered onto a profligate existence, constantly on the run to escape the consequences of debt, fraud, and high-society confidence tricks, for which she was eventually imprisoned. Since his mother's family had lived in Haiti during the eighteenth century, some have speculated that Gobineau was troubled by the thought of black ancestry, although no evidence of this has been found. Gobineau might also have had concerns about the ancestry of his wife Clémence Monnerot, who had been born in Martinique where it was common practice for French slave masters to take slave mistresses. This concern extended to their two daughters.

A loyal but not entirely obedient disciple of Broca, Paul Topinard led the Anthropology Society after Broca's death in 1880 until 1886, when a group of "materialists" with whom he quarreled succeeded in ousting him and replacing him with one of their own. This sequence provoked a wave of resignations that continued as the Society degenerated into factions and declined in influence and membership. In 1889, Topinard, planning a countermove against the materialists, solicited the assistance of the Director of Higher Education, on whom the Society depended. His maneuver was discovered and Topinard was dismissed from his chair at the Paris School of Anthropology.[71] He continued to advocate for what the Anthropology Society had, under Broca's and his leadership, described as "the broadest sense of anthropology."

Always less invested than Broca in the accumulation of statistical evidence, and distrustful of the direction in which Broca led the Society, Topinard wanted to view the human community as a single complex group comprised of a number of "types." Early in his career, he struggled to find ways of reconciling his fidelity to Broca with his suspicions about the whole enterprise. He began his 1876 tome *L'Anthropologie* with a complaint about the endemic imprecision of the available terms and the resulting impossibility of giving to each category the proper name. "That which is *family* in one [system]," he wrote, "becomes *order* in another; that which is *genus* becomes *species*; and vice versa." Noting that "the divisions and subdivisions of the human family are designated in current language by the name of *races*," he questioned the term itself: "Have these races the value of species, of varieties, or even of genera? That is the question." The introduction of "the human family" represented a cunning work-around for it had the effect of placing all humans in a single uncontested group. He concluded his long book by both affirming Broca's identification of races as species and blunting the effect of this identification by subsuming all species into one large category: "To sum up: The HUMAN FAMILY, the first of the ORDER of Primates, is composed of SPECIES, or fundamental human races."[72]

Later in life Topinard struck out in a more decisively independent direction, openly rejecting Broca's polygenism and the conviction that whites and blacks might be thought of as different species. In his 1885 *Elements of General Anthropology*, written five years after Broca's death, he separated himself from Broca by implication, associating his mentor with his fellow polygenists Nott and Gliddon, whom he described not as "distinguished" or "eminent," as Broca had, but as "*les esclavagistes*" (slavers).[73] More consequentially, in 1892, Topinard raised the possibility that what are now called races were merely the consequences of long isolation, environmental influences, and intermixture, and more radically that race itself was not an object for scientific research at all. It was, he said in a formulation nearly identical with statements made by Firmin, undiscoverable by any empirical means, and was in reality "only a subjective notion, a conception within our minds."[74]

Clémence Royer, a self-educated schoolteacher, republican, anticlericalist, and feminist, had been given the rights to translate *On the Origin of Species* by Darwin himself. The translation appeared in 1862 with a long preface in which Royer expounded many of her own views, not all of which were derived from or shared by Darwin, who referred to her in a June 1862 letter as "one of the cleverest & oddest women in Europe."[75] Most of the criticism of her translation attributed its defects to her sex, but Renan described her as "almost a man of genius."[76] In 1870, Royer became the first woman to be elected to a scientific society when she was admitted to the Société d'Anthropologie. After the death of her longtime partner in 1885, she supported herself and her son by giving lectures. She died in 1902, having lived the final twelve years of her life in a retirement home. A street in the first arrondissement in Paris is named in her honor.

CHAPTER FIVE

The Fall and Rise of Race

Races exist; it is undeniable; our intelligence comprehends them; our minds see them; our labors separate them out. If in thought we suppress the intermixtures of peoples, their interbreedings, in a flash we see them stand forth—simple, inevitable. . . . But nowhere can we put our fingers upon them.

—PAUL TOPINARD, *ELEMENTS OF GENERAL ANTHROPOLOGY* (1885)

FIRMIN HAD COME to Paris at the right time, just as the influence of Broca was declining, bringing down with it the revolutionary conviction as well as the prestige of the Anthropology Society. Firmin had somehow sensed the point of vulnerability in the Society's emphasis on anatomical and especially craniological evidence, and had applied his formidable pressure precisely there. It was not, however, an effective intervention. There is no indication in the Society's records that he had any influence within the Society or that anyone in it even read his book; an authoritative history of the institution published in 2004 makes no mention of it, or him, at all.[1]

One possible reason for this disregard may have been that his critique was unwelcome, but another may have been that it was redundant. By the time Firmin published his work in 1885, the Society had fallen into acrimonious confusion about the fundamental question of whether anthropology was part of the natural or the human sciences. With the new discipline weakened by internecine conflicts and uncertainty about its fundamental mission, intellectual leadership had passed to other disciplines more certain of themselves. So compromised was anthropology that, as one scholar notes, it was unable even to be an effective ally in the service of the colonial or imperial causes that were, according to its critics, its *raison d'être*.[2]

Broca had staked his and the Society's reputation on the possibility that meticulous documentation of anatomical and craniological measurements could provide a scientific solution to the problem of human variation. As we have seen, Broca was content to call this the problem of race in part because the ambiguous term left open the question of whether groups should be considered variations on a single model or separate and distinct species. Monogenists preferred to think of them as variants while polygenists thought of them as species; the advantage of race, Broca had said, was that its equivocality left open the larger question pending further research. In the years after Broca's death in 1880, the entire project, with no sponsor of comparable stature or charisma, came to seem increasingly dubious. The theory of race, so earnestly cultivated by scientists and intellectuals in Europe and the United States for over a century, was unraveling.

Defections commenced.

Jean-Louis Armand de Quatrefages de Bréau, one of the first and most distinguished members of the Anthropology Society (and sponsor of Royer's membership), had begun his career as a zoologist before accepting an appointment in 1855 to the chair of anthropology and ethnography at the Muséum national d'Histoire naturelle in Paris. Eager to apply zoological methods to the human species and to undertake a study of the crania of human fossils that were being unearthed, Quatrefages was nevertheless skeptical of Broca and openly contemptuous of Nott and especially of Agassiz, whose arguments he subjected to an almost cruelly protracted demolition in *The Human Species* (1877).[3] The lack of definition that Broca saw as an advantage Quatrefages saw as an impediment. The blurring of any distinction between race and species represented, Quatrefages said, a determined confusion, "a want of precision in general ideas," and even a suppression of basic facts.[4]

Quatrefages saw Nott's polygenism as an imposter of science that pretended to neutrality on social or moral questions while relentlessly generating statistics and distinctions that in fact supported the indefensible practice of human slavery. The fully intended effect of the attempts by Nott and others to bring the authority of the Bible into question and to promote polygenism was, Quatrefages said, to provide some ground on which proslavery advocates could argue by insisting that science had identified "radical differences" between the races that precluded any kind of social or political equality.[5] Quatrefages was far more professionally qualified than his opponents to speak about the fertility criterion of species identity and was not reluctant to point to the prolific offspring of the

FIGURE 5.1. Jean-Louis-Armand de Quatrefages de Bréau. Public domain. Source: Wellcome Collection, London.

victims of slaveowner rape as evidence that races were not species by the commonly accepted definition of that term.

Quatrefages made his positive argument for monogenesis in a central chapter of *The Human Species* on "Formation of Human Races under the Sole Influence of Conditions of Life and Heredity," in which he posited an original race whose members "at first. . . . differed from each other only in individual features," but who, under varying "conditions of life" diversified into what are called races. The differences in some cases might, he said, have become so marked that members of one group came to regard members of another with the seemingly natural or instinctive revulsion, distrust, or hostility that was often cited as a mark of species differences. In such cases, "the *acquired nature* is, so to speak, welded to the original nature of the being." But all human groups have acquired some "nature" or other; after the glacial era, "the earth has no longer had *autochthones*, but has only been peopled by *colonists*."[6]

Indeed, *The Human Species* represents perhaps the most direct indictment by a recognized scientist not just of the defeated slave-owning Confederacy but of the globally colonizing "White" who, "impelled by his restless ardour, has invaded the world, and is every day multiplying his conquests and colonies. It is he," Quatrefages continued, "who has searched out the home of the coloured races, and who everywhere mingles his blood with their own. Almost all the half-breed populations recognize him as their father."[7] Groups formed by this means were, Quatrefages believed, neither permanent nor rankable, proof of which was that in the course of human history ascendancy had passed from one race to another, while the groups themselves exhibited a good deal of "plasticity."

In no group was this capacity for self-alteration more immediately visible than in the North American Negro, who had by all accounts including that of Nott become lighter in color, more evidently "intelligent," and less starkly different from those of European descent than their ancestors had been. Signs of plasticity were evident to others as well. When the geologist Charles Lyell visited a church in Savannah, Georgia, Quatrefages noted, "he remarked that the odour so characteristic of the race was scarcely appreciable." The races in North America were inclining each toward the other, so that eventually, "whatever be their origin, all the descendants of Whites or of Negroes who have immigrated to America will become Red-skins."[8]

This positive case was based on the evidence of the eye, or nose. But *The Human Species* ended with a section on "present human races" that amounted to a negative case against those, beginning with Broca, who presumed to identify species differences in the skin, hair, proportions, stature, skeletal structure, skull shape, or labia minora, none of which, Quatrefages insisted, could bear the weight of the argument. In every case where measurements or statistics had been proposed as evidence of some inherent "racial" feature, Quatrefages argued for conditions of life and heredity as the real causes.

The final two chapters of *The Human Species*, on "Moral Characters" and "Religious Characters," constitute a frontal assault on European vanity, pride, condescension, and *amour-propre*. Are "savages" unacquainted with the "moral sense"? Conventional wisdom says yes, and points to incidents of non-European savagery as proof. In response, Quatrefages recalls "the orgies of Greece and Rome," as well as "the terrible revelations which from time to time are made in the police courts of our proudest capitals" as

FIGURE 5.2. Rudolf Virchow. Public domain. Source: Wikimedia Commons.

evidence that the "*savage element* still exists in the most *civilized nations*." And as for Africa, "it seems to me that the two words, *trade* and *slavery*, are quite sufficient to prevent a European from boasting too loudly of the morality of his race."[9]

Like Quatrefages, Rudolf Virchow saw the entire massif of the science of racial difference as the result of a motivated imprecision that deliberately confused race with species. Even the terms for human variation were misleading because they contained embedded answers to the question of race that must, Virchow insisted, remain open. Scientific terminology had become degraded to the point where it was, as he said in an 1896 essay on "Heredity and the Formation of Races," "not sharp enough to maintain always exactly the same meaning for race and stock. The concept of race has gradually become so diffuse that whenever it suits the expounder the designation can also be extended to a stock."[10] Some expounders, he implied, used the concept of race when they wanted to draw a sharp line between groups, often in pursuit of nonscientific ends.

In defending the integrity of science, Virchow spoke with authority. A figure of immense and general celebrity, he was the most distinguished physician of his day, recognized for pioneering work in cell biology, epidemiology, and pathology—fields he thought provided a model for understanding social and political processes. In addition to his work as a physician, he was one of the leading anthropologists of his day, having founded both the German Anthropology Association and the Berlin Society for Anthropology. He had explored prehistoric sites in Asia Minor, the Caucasus, and Egypt, and had examined the first Neanderthal fossils. A man of strong convictions, Virchow regarded Darwinism as a suggestive but unconfirmed hypothesis and Darwin himself as an "ignoramus." No skeptic about the value of measurements and statistics, Virchow analyzed the hair, skin, and eye color of nearly seven million schoolchildren in an attempt to differentiate Jews and Aryans; when his efforts failed to yield any patterns that would constitute an argument for distinct racial differences, he announced that neither a Jewish nor a German—nor an "Aryan"—race existed.[11]

Similarly unimpressed by efforts to identify particular skulls as "Turanian," "Slavic," or "Teutonic," Virchow opposed any attempt to connect national identity to race. There was, he insisted, simply no way to tell from a skull which race its previous occupant/possessor had belonged to. With German political culture under Bismarck heading in the opposite direction, toward a confident racialism, Virchow became so controversial that Bismarck himself challenged him to a duel, which Virchow prudently declined. But he continued to argue against the false exactitudes of racial science, regarding craniology with particular contempt as a source of support for "Nordic mysticism."[12] Having no basis in skin color, hair, skeleton, or skull, the entire theory of race was, he declared, simply null. Theories had "out-distanced the facts," and in the absence of new evidence, he declared, further research would be wasted.[13]

Among those coming to the same conclusion was Joseph Deniker, a Russian-born polymath with a distinguished record of achievement in linguistics, comparative anatomy, geography, and embryology who was, at the end of the century, a member of the Anthropology Society of Paris and the librarian of the Muséum national d'Histoire naturelle, where Quatrefages was chair of anthropology and ethnology. His magisterial *The Races of Man* [*Les races et les peuples de la terre*] (1900) reflected his work in all these areas, but particularly in geography.[14] The "racial map" of Europe he developed for this volume provided a startlingly precise image of the distribution of six "primary races" and four "secondary races" over the European continent, including the "Nordic" race in northern Germany, Scandinavia, and the British Isles.

Deniker was also an astonishingly industrious craniologist, having been responsible for establishing the dolichocephalic, mesocephalic, and

FIGURE 5.3. Joseph Deniker. Public domain. Source: Gallica.

brachycephalic measurements for a dizzying number of categories and subcategories, some of which he introduced or to which he gave new names. In all, he managed to identify twenty-nine races organized into seventeen groups. Deniker's unquestioned command of the vast body of relevant facts relating to race made him an important voice, and at over six hundred pages with over one hundred and seventy illustrations, *The Races of Man* was clearly intended as a standard text, the most exhaustive account of the racial composition of the human species ever undertaken. But so great was the sheer mass of statistics and data—all those millions of skulls measured, some in hundreds of ways, the figures having been duly reported by physicians all over the world—that it overwhelmed even his capacity to organize and analyze them.[15] Just as discouraging to

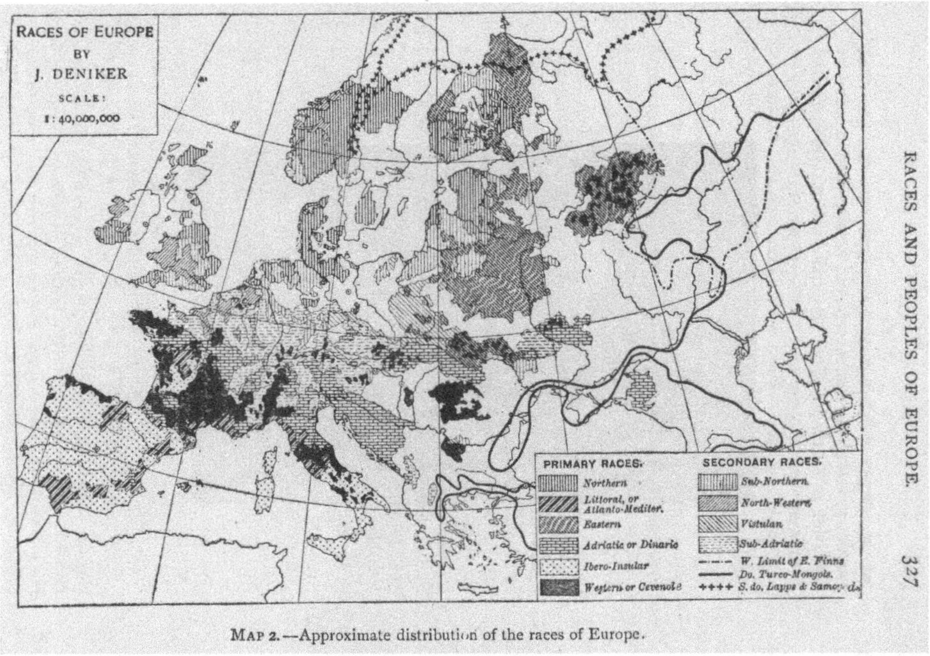

FIGURE 5.4. Map of the races of Europe, from Joseph Deniker, *The Races of Man* (1900), 327. Source: Wikimedia Commons.

the researcher looking for evidence of categorical differences, the average weight for European and Negro brains turned out to be almost identical. In the end, Deniker joined Topinard, Quatrefages, and Virchow in advocating for the abandonment of race as an object of scientific research.

In fact, this conclusion comes not at the end but at the very beginning of his book. Titled "Ethnic Groups and Zoological Species," Deniker's introductory essay both establishes the subject and despairs of an adequate treatment of it. "A considerable diversity of opinion" among leading scientists has, he notes, resulted in a stubborn imprecision in the very terms that ought to be providing clarification. People, nation, tribe, race, species—terms for "primordial divisions"—are "very vaguely indicated." "Type" might be an improvement over some other possibilities, but the concept is far more applicable to animals than to humans, who group together in clusters that have no zoological status.[16] And by what criterion might one distinguish a species from a subspecies? He could identify none—none of the words made sense to him.

Deniker proposed as a more defensible alternative the term "*ethnic groups* formed by virtue of community of language, religion, social institutions, etc." This was a bit clearer than the others, and had the added benefit of removing the entire discussion from the zoological realm of "man, considered as an animal." This removal was entirely appropriate, Deniker felt, because most people present the scientist with "only a confused mixture of characters," with the confusion increasing with every step up the ladder of civilization, where factors of mobility and choice increased the number of "somatological units" in the individual. For the scientist trying to slot human groups into natural categories, the whole situation was hopeless. "It seems to me," Deniker said, "that where the genus *Homo* is concerned, one can neither speak of the 'species,' the 'variety,' nor the 'race'" except in the very broadest sense, and then with great reluctance and only out of grudging respect for convention.[17] While Deniker continued to describe, categorize, and map, his immense work became a monument to concepts he could no longer defend. As Walter Scheidt wrote of Deniker, Quatrefages, Topinard, and others in 1924, "Hereby anthropology reaches a point, through its anthropometric technique, where its most notable representatives urged that the solving of its principal problem, that of race, be abandoned."[18]

By the end of the nineteenth century, faith in the evidence, the scientists, the institutes, the methods, and above all the concept was rapidly waning. Darwinism, once the object of widespread skepticism as an unproven and unprovable theory, was becoming an irresistible force, and the markers of race that had supported so much theorizing were coming to seem secondary and derivative phenomena when compared to the multi-million-year transformations Darwin had described. Darwin's very limited interest in the "so-called races of man" reflected his sense of the irreducible ambiguity and even dubiety of the concept.[19] As he pointed out, no consensus had ever been established on the number of races, with authorities arguing for numbers ranging from one (the human) to sixty-three.

The number was uncertain because it was unclear what constituted a race. At one time or another, Franks, Gauls, Caucasians, Americans, Lapps, Celts, Negroes, Teutons, Aryans, Border- Scots, Nordics, Irish, Jews, Scythians, Tartars, Hottentots, Malays, Slavs, and Europeans had been described as separate races. The site of race—skin, hair, bones, eye color, ear form, nose type, stature, blood, brain tissue, sweat, genitals, social customs, language, geography, skull size or shape, souls, smell, body lice, or semen, all of which had been seriously proposed at some point—remained undetermined, a series of dry wells sunk by some of the leading scientific minds of the century. A manifest reality to the unscientific eye, race could not be discovered by empirical methods and was revealing itself to some of those who had dedicated years of their lives to the effort to be no more than a statistics-rich muddle that had become embedded in the language.

Before the futility of the project was realized, however, the concept of race had been worked hard, serving as a screen or proxy for other issues that were difficult to address directly and did not respond to research—the Enlightenment legacy, the ultimate origins of humankind, the status of revealed religion, the grounds of national identity, the legitimacy of human rights, the place of mankind in the natural order, the obligations of the more powerful or culturally advanced nations toward others, the basis in nature for certain historical events or social arrangements, the ultimate cause and meaning of human diversity, and the human condition more broadly.

Eminent minds had thought it possible that a scientific understanding of race would provide the key to these diverse kingdoms. Reflecting on a century of earnest scientific labor, Henry Adams wrote of race that "history offered a feeble and delusive smile at the sound of the word; evolutionists and ethnologists dispute its very existence; no one knew what to make of it; yet, without the clue, history was a fairy tale."[20] The thronging background presence of issues of existential import had given to debates about race a dramatic intensity that few scientific debates then or now possess, but at century's end it was race rather than history that was coming to seem a fairy tale.

Among those hastening the demise of race as an object of scientific inquiry was a flamboyant young German-American student of Virchow, Franz Boas, who was just then beginning what would be a long campaign to get his fellow anthropologists to give up the whole subject, for the brutally simple reason that, as he said in 1936, "nobody can tell us what a race is."[21] As a young man, Boas had delivered a blow to the head of craniology, establishing that the various measurements of body and skull that were at that time still respected by some as objective evidence of racial categories were not fixed but were influenced by health and nutrition. His extensive series of measurements on the heads of the children and grandchildren of immigrants to America proved that in as few as two generations, the skulls of immigrant families more closely resembled those of their new neighbors than those living in their former ancestral homes.[22] The hard evidence thought to be provided by skulls was no harder than that provided by anything else. Indeed, the concept of race was taking on the aspect of a skull, an inert enclosure capable of containing many thoughts, not all of which were in agreement.

Boas's work spoke to many who had become convinced that the study of race was not only a failed but a corrupted enterprise, a taxonomic convenience proposed in the eighteenth century that, having been

FIGURE 5.5. Franz Boas, ca. 1915. Public domain. Source: Wikimedia Commons.

misconstrued as a fact of nature, had become a source of evil: an unresisting and often willing accomplice to projects of imperial domination or exploitation, an ideological prop for American slavery and racial persecution, and a source of scientific authority for various forms of racist biopolitics including the fascistic German race-state.[23]

Having served many ends over the course of the preceding century, the concept of race now stood accused of being *all ends*, none of them scientific. Having taken up race as a way of pursuing common-man questions—Is humankind one or many? Are races species? Where did races come from and how many are there? What place does humanity occupy in the natural world? In what ways are races similar or different? Do some races have more intellect, sensuality, or courage than others? Are pure races superior to mixed races?—many scientists were beginning to think that an increasingly professionalized scientific discourse was not only unable to address them but that it could continue to raise them only at the cost of its own credibility. If science had discredited the notion of race, they felt, then a continued inquiry into race could only discredit science. The doubts about the concept voiced by Herder, Darwin, Douglass, and others now seemed prescient and compelling to those who had formerly found ways and reasons to set them aside. With the most respected scientists in sudden agreement, the research project that had occupied some of the best scientific minds in the world for over a century collapsed.

In retrospect, 1900 was a watershed year in the history of the race concept. Not only had leading anthropologists gone public with their conviction that the problem of race could never be solved by empirical science, but the entire issue of inherited traits was placed on a new footing by the rediscovery and verification that year of some of the experimental findings of Gregor Mendel several decades earlier. Working with mice, bees, and plants, Mendel had discovered that heredity was a function not of the blending or averaging of the characteristics of the two parents but of the independent and discontinuous assortment and recombination of genetic units that were common to all humans. "Racial" features such as skin color, hair texture, or skull shapes were not genetic in the Mendelian sense at all, but were rather morphological features, reliably transmitted in geographically or socially isolated populations but rapidly diluted in mixed or mobile groups. If the inquiry into race was to continue after Mendel, it had to invent new ways of thinking.

The determination and ingenuity of some scientists in sustaining the subject as a topic for research in the first four decades of the twentieth century cannot be overstated. Many of those who did so acknowledged that Topinard's statement that we cannot "put our fingers" on races was a problem, but chose to emphasize instead the fact that, as Topinard had said, "our intelligence comprehends them," and proceeded to devise new ways of continuing the work. With respected researchers leaving the field in varying states of frustration, despair, or disgust, the term race, now part of the common vocabulary, was adopted by a swarm of ideologues promoting various ends—eugenicists, master race polemicists, anti-immigration fearmongers, and degeneration-alarmists, some of whom enjoyed professional esteem and were associated with established universities or laboratories. Their number included Houston Stewart Chamberlain, Roland B. Dixon, G. Stanley Hall, Earnest A. Hooton, Karl Pearson, Charles Davenport, H. H. Goddard, Madison Grant, Cesare Lombroso, Max Nordau, and Lothrop Stoddard, to mention just a few of those whose names circulated in the Anglo-American context in the first years of the twentieth century.

Within the more traditional precincts of academia, the inquiry was kept alive by diverse means. Scholars and scientists in the increasingly professionalized fields of anthropology, paleontology, and biology proliferated terms, distinctions, considerations, and criteria that had the effect of preserving the concept even after confidence in the ability of science to discover positive evidence of it had evaporated. New information derived from fossils and prehistoric migratory patterns, along with new terminology for categorizing bodies (homo famobrachycephalus, homo dinaricus, the "respiratory type," microsplanchnic and mesosplanchnic [with subdivisons], hypomorph and mesomorph, platyrrhine, mesoprosopic, hypsicephaly) seemed to offer promising approaches to an old project that had run aground. New measurable physical features were proposed as possible indicators of racial identity including the postglenoid tubercle, prognathism, sutural patterns, papillary patterns in fingers, the interpupillary distance, the pharyngeal fossa, the supracondyloid process, and the hormonistic index. With new entities emerging all the time, it was possible for some to feel that, uncertainties and pettifoggery aside, much good and honorable work remained to be done.

At the same time, statements that races "exist only as fictions in our brains" became increasingly commonplace in the scholarly literature, as were confessions that the question had become "somewhat confused and uncertain," that scientists had "a lack of unanimity in the matter of what is a race," and that the entire subject was beset by "an unavoidable looseness in the concepts." While some suggested that "it is better not to use the term race at all in view of its painful misapplication in political discussions," others held out hope that the current confusion would eventually give way to the advance of knowledge.[24]

Expressions of caution or warnings about reaching premature conclusions were offered as assurances that science was being responsible, methodical, and prudent even as the concept was being conscripted into political or ideological struggles pitting the advanced peoples of the world against the "darker races," sometimes delicately characterized as "the types less dowered."[25] The possibility of deep and unalterable divisions between what Topinard called "simple general types" within the human species lent a sense of ominous consequence to the IQ tests developed at the beginning of the century and a sense of clarity and purpose to programs such as Karl Pearson's 1901 call for the "Aryan" race to elevate the Western "stock" by eliminating inferior races through war and conquest.[26] For such projects, race was not an object of study, not part of an inquiry into human variation, but a premise, a vernacular concept whose generality and imprecision shielded it from objections.

The awkward coexistence of confident assertion and persistent doubt was placed on public display in the landmark 1911 *Encyclopedia Britannica* essay on "Negro." Trailing a bit behind the scientific consensus just described, this long and justly infamous article retails a host of unreflected convictions inherited from the science of the previous century—that "the negro would appear to stand on a lower evolutionary plane than the white man"; that a relation between the negro and "the highest anthropoids" is established through arm length, zygomatic arches, and a depressed nose; that the negro is mentally inferior to the white especially in adulthood, when childhood vivacity is replaced by "a sort of lethargy" attributable to the "premature closing of the cranial sutures" and the adult's obsession with sex; and that the adult negro has remarkably acute senses but a mental constitution "very similar to that of a child." This unembarrassed list concludes with the stunning admission that while census enumerators have always distinguished between "the two main races of white and negroes . . . they have never been given a definition or criterion of race."

This essay was in fact coauthored by a British ethnographer and a statistician working at the United States Census Bureau. In the next edition of the encyclopedia, published eleven years later, the same entry—this time authored by the statistician alone—scrapped ethnography altogether in favor of demographics based on American census figures. It is reasonable to suspect that embarrassment finally kicked in for the editors when they saw how closely their 1911 entry conformed to the premises of the Ku Klux Klan, which was revived in the first years of the twentieth century.[27]

During the first decades of the new century, the context created by world events pressured the scientific inquiry in ways that could not be denied. The explicitly racial basis of Germany's conduct of the second World War made it all but impossible to claim scientific neutrality on, for example, the issues of plural origins, blood group data, or hierarchies of traits. In 1939, Ruth Benedict, a student of Boas, identified and attacked "racism" as "the 'ism' of the modern world," citing the racial ideology of the Third Reich as the leading instance.[28] For Benedict, racism was but the contemporary form of an immemorial phenomenon, and "to understand race conflict we need fundamentally to understand *conflict* and not *race*."[29] She defended the study of race as a legitimate scientific inquiry into the genetic relations among human populations, but just a few years later, even this position came to seem untenable. In 1942, Ashley Montagu condemned not perversions of the concept but the concept itself as "man's most dangerous myth."[30] In the years after the second World War, the combined force of various antiracist and anticolonialist movements, added to forcefully articulated scientific arguments, put the final nail in the coffin of the idea that race was discoverable by scientific means.

This conclusion represented a considerable achievement. Science, in increasingly professional and self-aware forms, had been *primarily* responsible for developing the concept of race over the course of the nineteenth century. But it was *entirely* responsible for discrediting the idea that categories of humanity were facts of nature. The recognition that race is invented, constructed, or mythical, which had been strongly suspected by serious thinkers from the beginning, might never have been established by nonscientific means as anything more than a dubious hypothesis that contradicted the powerfully persuasive factors of plain observation and everyday experience—what Topinard referred to as "our intelligence"—without the endorsement of the scientific establishment. In a restricted sense, then, we can be grateful to those who steered the inquiry into race onto the terrain of hybridity studies or biometrics, two areas in which the evidence for biological categories was always going to be negligible, for by doing so they sealed its fate, ensuring that the research project focusing on race would eventually fall of its own weight.

If we owe to science the invention of race, then, we also owe to science the discovery that race is an invention, or to be more precise, that there is nothing in the physical constitution of human beings that answers to the concept of race. By abandoning race as a problem to be addressed by biological research, leading scientists sought to vindicate science itself not by producing positive knowledge of reality but by acknowledging their failure to do so and, just as important, by acknowledging the extraordinary success of science in producing positive knowledge of a nonexistent object.

The collapse of the empirical inquiry into the nature of race—the depth and kind of the differences among human populations and the mechanisms that might have produced them—constituted a triumph of the institution over the individual. Throughout the nineteenth century, this inquiry was conducted by sometimes exceptionally gifted but fallible people constrained by sometimes unsuspected limitations of knowledge, perspective, and sympathy. Their various failures to develop a credible and consistent account of race weigh on their individual reputations, but can be understood as part of a collective success. If the science of race could at times have been justly accused of complicity with practices of exploitation or oppression, the abandonment by leading scientists of race as a topic for scientific inquiry at the end of the nineteenth century suggests a different story, about the awakening of science to its own imperatives, its own way of knowing, and its own conclusions. Included in this awakening, of course, was the chastened realization that many scientists and intellectuals who had enjoyed general respect and even celebrity had committed themselves to premises that were now understood to be corrupt, methods to be deeply flawed, and conclusions to be erroneous.

The delegitimization by leading scientists of race as an object of empirical research did not eliminate "racial" discrimination or injustice, but it did what science could do, de-realizing the concept so that it was supported only by other forces whose nonscientific determinants were immediately visible to many if not to all. It is only as a result of the (disorderly, faltering, error-ridden, all-too-human) scientific inquiry into race that we are now able think of race as an unreal reality—a heuristic, a convenience, an artifact, a "mental tool," a part of the "built information environment."[31] The recognition that the topic of race is a subject for social psychology rather than biology remains the most decisive and incontrovertible result of the entire project, and the one that has become part of received wisdom, even of common sense—which, however, credits itself with the discovery and holds science responsible for the error that science in fact exposed.

As the scientific community was going through this dramatic process of self-scrutiny and self-assertion, another kind of discourse, conducted on different terms altogether, was gaining strength. Rising from the ashes of a failed attempt to discover a category of nature, the concept of race was taken over by cultural forces that defined it as a substantially dematerialized principle of social identity. In 1897—three years before the Pan-African Congress organized by Firmin—W.E.B. Du Bois delivered an address to the American Negro Academy titled "The Conservation of Races" in which he called for the abandonment not of race but of science.

Dismayed by the prospect that the Negro race with which he identified (both parents were mixed-race) and even race itself might be declared by science to be mere mirages, Du Bois also saw an opportunity created by the emerging scientific consensus to argue for a nonscientific understanding of race as the product of "common blood, descent and physical peculiarities."[32] Frederick Douglass and Anténor Firmin rejected race as an obstacle to equality, but Du Bois saw a value in it that they did not. A reinvigorated concept of race as a principle of identity achieved through organization and

FIGURE 5.6. W.E.B. Du Bois, 1907. Public domain. Source: Wikimedia Commons.

solidarity was, he said, required in order for Negroes to complete the work of liberation to which these men had committed their lives.

Other groups, he pointed out, had made their mark—the English had contributed constitutional liberty, the Germans science and philosophy, and so forth—but the Negro had not yet given to the world "the full, complete Negro message of the whole Negro race." It was therefore the duty of American Negroes in particular to "maintain their race identity until this mission of the Negro people is accomplished, and the ideal of human brotherhood has become a practical possibility."[33] There was, he argued, a worthy end to which race was an invaluable means, and as long as the end remained unrealized, the concept of race had to be conserved. If science no longer supported the concept, black people themselves must take over the curatorial responsibilities, beginning by establishing a more productive understanding based on values and purposes that could not be formulated in scientific terms.

What, then, is race? According to Du Bois, race is a principle of elective affinity that is still grounded in reality—a "vast family of human beings, generally of common blood and language, always of common history, traditions and impulses, who are both voluntarily and involuntarily striving together for the accomplishment of certain more or less vividly conceived ideals of life." Du Bois understood that race is necessarily both a biological concept and more than a biological concept. The "real distinctions" on which race is based include, he said, the "grosser physical differences" evident to the eye, but also comprehend "spiritual, psychical, differences—undoubtedly based on the physical, but infinitely transcending them." Inaccessible to science, these "subtle, delicate and elusive" qualities were nevertheless crucial to the identity, and thus to the advancement, of the Negro people.[34]

This was as close as Du Bois ever got to clarity on the definition of race.[35]

In a widely read 1986 essay called "The Uncompleted Argument: Du Bois and the Illusion of Race," Kwame Anthony Appiah argued that even if the ideal of human brotherhood had not yet been achieved, the time for conserving the race concept had passed. Nor was the concept itself as Du Bois had developed it even coherent, with its confusion of the biological and the cultural—common blood on the one hand and common impulses and strivings on the other. The insistently familial metaphors Du Bois used to describe racial identity such as "descent," "heritage," "kinship," and "ancestors," as well as "fatherland" and "motherland," could not, Appiah insisted, apply to huge numbers of genetically unrelated people with vastly different experiences, languages, values, and cultural practices. Indeed, Appiah pointed out, Du Bois himself revised his answer to the question of race several times over the half century following his first essay, emphasizing now the biological, now the political, now the experiential, or now the spiritual. His failure to settle on a single comprehensive definition actually demonstrated the unworkability of his theory of race. "The truth," Appiah declared, "is that there are no races; there is nothing in the world that can do all we ask 'race' to do for us."[36]

When he wrote this, Appiah was already aware that his dismissal of race would not be persuasive to all, and concluded his essay on a note of vexation at the fact that, incoherence notwithstanding, Du Bois "is followed . . . by many in our culture today."[37] Indeed, some of those followers immediately attacked Appiah, who over time—buckled. Delivering the Du Bois Lectures at Harvard nearly thirty years after his first essay, he came

FIGURE 5.7. Kwame Anthony Appiah. Source: Berggruen Institute, Los Angeles.

to a different conclusion, declaring that if there is nothing in the world that can do what we ask race to do, then there is equal reason to say that the word is indispensable. This was in fact Du Bois's position, that race represents a valuable principle of grounded identity or belongingness that cannot be formalized under any other term.

And so, on consideration, a matured and chastened Appiah joined the ranks of his critics, giving a highly sympathetic reading of Du Bois on the utility of the race concept. "Notice," he said in a sly footnote to the published version of the lecture, "that I part ways here . . . with the author of 'The Uncompleted Argument: Du Bois and the Illusion of Race.'"[38]

One might wonder why or even whether Appiah truly changed his mind, or whether he simply distributed over two essays and three decades the ambivalence that Du Bois had compressed into a single shimmering statement. How, indeed, could one be unambivalent? The thought of natural boundaries or barriers between categories of human beings may be clarifying and empowering in some ways, but it is bleak and mysterious in others, and can be an obstacle as well as a path to human solidarity. Promoting a conflicted concept as both problem and solution, "The Conservation of Races" contains its own refutation, which a sympathetic reading will register and, in Appiah's two-stage case, repeat.

One conclusion that might be drawn from the history of the concept of race is that we need a better theory than those offered so far. Another, however, is that there is no comprehensive and coherent theory that reconciles the biological—the inherited "grosser physical differences" without which identity and ascription have no purchase on the world—with the impulses, strivings, aspirations, and experiences that constitute a sense of identity. As Darwin, Firmin, Douglass, Topinard, Quatrefages, Deniker, Virchow, Boas, and the younger Appiah recognized, the term race does not correspond to anything in nature; it is shaped, determined, and sustained not by the evidence that supports it but by the ends it can be made to serve.

Many of those ends represent interests or questions fundamental to an advanced intellectual culture. We want to understand the origin of the apparent differences among peoples. We want to know the place of humankind in the natural world. We want to understand the nature of our connection to, or disconnection from, others. We want to know whether some hidden system regulates the distribution of attributes among peoples. We want, in short, to understand the human condition. Other ends are fundamental to social life. We want to understand the differences between our group and others, and to know how to think of otherness and difference in general. We want to know who to love and trust and who to despise and fear. For over two centuries, race has provided a general heading for thinking on such subjects.

The kinds of thinking race has sponsored are, however, never simple and the blessings it confers never unmixed. The blurred or equivocal character of race, referring to both humankind in general and to groups within the human, has enabled people to discover in it some support for their favored conclusions, but it has also ensured that other conclusions can be supported as well. In fact, every statement made about race is shadowed by its negation. The argument that race is a bodily fact or category of nature is shadowed by the thought that it is only an artifact of the scientific, philosophical, or ideological imagination, and that it can never be anything other than asserted, ascribed, or imputed. But the opposite argument, that race is a social construction or an ideological weapon, is shadowed by the often unstated premise that race has a biological basis, a premise that sometimes becomes explicit in response to incidents in which a certain racial identity is claimed by one who lacks the genealogy to support the claim.[39]

In his biography of Du Bois, David Levering Lewis recounts an incident that demonstrates the point. At a meeting of the American Sociological Association in Madison, Wisconsin ten years after Du Bois published his essay, he was invited to reply to a paper given by Alfred Holt Stone, president of the Mississippi Historical Society and son of a Confederate soldier. As it happened, Stone agreed with Du Bois on the necessity of conserving the concept of race, but differed as to the reason, saying that "the superiority of race cannot be preserved without pride of blood and an uncompromising attitude toward the lower races." As Lewis describes the incident, Du Bois, seated in the audience, might have felt that his own idea "had returned to haunt him in a demented form."[40] But Holt's form seemed plausible enough to many, even perhaps most Americans, who were at that time passing Jim Crow laws throughout the country premised on the conservation of races.

Every Du Bois, it seems, has his Stone, every Blumenbach his Meiners, every Morton his Nott, every Darwin his Royer, every Gobineau his Hotz, every Firmin his . . . Firmin.

Du Bois described the "double consciousness" that affects every American of African descent. But in a more general sense, a deep equivocality haunts the concept of race itself, which inspires a sense of solidarity and identity by positing fissures and divisions. This doubleness can be seen everywhere in the discourse on race, even, for example, in the secondary terms that gather around the subject in the United States. The term Black with a capital *B* is thought by many to foster a sense of dignity among the members of the designated group, but the same term was used by John C. Calhoun as a way of naturalizing and concretizing the category of the justly enslaveable. The phrase "persons of color" appears in many Jim Crow laws. "Diversity" was the favored term of those including Morton and especially Nott, whose sympathy with Calhoun's political goals took the form of arguing that Blacks were so very diverse that they constituted a separate species.

In one sense humankind is homogeneous and indivisible; in another sense there are differences within the category of the human. In no sense is humankind divided into natural and unalterable, much less God-created categories. While it is difficult to think of a counterfactual world in which the ends served by race do not exist, it is possible to imagine a world in which the specific concept of race had never emerged, or in which, having served the time-limited purpose Du Bois identified, it just withered away, a superannuated relic of an earlier time.

And so the question arises: Is the conservation of race still necessary? Is it the best we can do? Or, given the accelerating inexorability of mixture, should race, having underwritten over two centuries of thought and practice, now be recognized as a kind of skull, a remnant of an era marked by suffering, indignity, conflict, arrogance, violence, and confusion, a once-living concept now surviving in the cadaverous forms of professorial subtleties, outmoded bigotries, and identity politics with no more than a tactical justification? At the conclusion of his 2014 book *The Myth of Race: The Troubling Persistence of an Unscientific Idea*, Robert Wald Sussman spoke for a contemporary consensus, denouncing the entire scientific—or as he insisted, pseudoscientific—inquiry into race, but asserting that "the cultural reality of race is real."[41] He argued, in other words, not against race but against a false view of race. It is a compelling argument—it has swept the field—but is it sufficiently imaginative? After all, what culture has done it can undo, or do differently. We might now ask whether race, having served so many ends, should now itself come to an end. Having risen and fallen and risen again, should race now fall once more, this time for good?[42]

We have inherited from our troubled history an obligation to answer this question.

NOTES

Chapter One: How Race Became a Problem

1. A student of the history of the race concept does not lack for resources. For a collection of primary texts from the eighteenth to the mid-twentieth centuries, see Earl Count, ed., *This Is Race: An Anthology Selected from the International Literature on the Races of Man* (New York: Henry Schuman, 1950). For an edited selection of texts from 1684 to 1900, see Geoffrey Galt Harpham, *Theories of Race*, theoriesofrace.com. This website can also be accessed through the website created by Henry Louis Gates Jr. and Andrew Curran in connection with their book *Who's Black and Why?* (Cambridge, MA: Harvard University Press, 2022), https://www.whoisblackandwhy.com/authors-who-is-black-and-why. See also Robert Bernasconi and Tommy Lee Lott, *The Idea of Race* (Indianapolis: Hackett, 2000); H. F. Augstein, ed., *Race: The Origins of an Idea, 1760–1850* (Bristol: Thoemmes Press, 1996); and Michael Biddiss, ed., *Images of Race* (Leicester: Leicester University Press, 1979). In addition to works cited in individual notes in this chapter, I have found the following particularly useful. One of the first comprehensive accounts was Ivan Hannaford, *Race: The History of an Idea in the West* (Washington, DC: Woodrow Wilson Center Press, 1996). Of unusual interest for a relatively brief and focused essay is Bronwen Douglas, "Climate to Crania: Science and the Racialization of Human Difference," in *Foreign Bodies: Oceania and the Science of Race 1750–1940*, ed. Bronwen Douglas and Chris Ballard (Canberra: Australian National University Press, 2008), 33–98. Among Robert Bernasconi's many contributions to a critical philosophy of race are *Race* (Oxford: Blackwell, 2001), and *Critical Philosophy of Race* (New York: Oxford University Press, 2023). For religion and race, see Colin Kidd, *The Forging of Races: Race and Scripture in the Protestant Atlantic World, 1600–2000* (Cambridge: Cambridge University Press, 2006). George W. Stocking Jr. has written two excellent books on race and anthropology: *Victorian Anthropology* (New York: Free Press, 1987), and *Race, Culture, and Evolution: Essays in the History of Anthropology* (Chicago: University of Chicago Press, 1982). See also Michael Banton and Jonathan Harwood, *The Race Concept* (Westport, CT: Praeger, 1975), and Nancy Stepan, *The Idea of Race in Science: Great Britain, 1800–1960* (Houndmills, UK: Macmillan, 1982).

2. Douglas, "Climate to Crania," 34–35.

3. Ania Loomba and Jonathan Burton, introduction to *Race in Early Modern England: A Documentary Companion*, ed. Ania Loomba and Jonathan Burton (New York: Palgrave Macmillan, 2007), 1–36. This book consists primarily of primary documents that bear on the question of the premodernity of race. Among the relevant texts are Adam Hochman, "Is 'Race' Modern? Disambiguating the Question," *Du Bois Review: Social Science Research on Race* 1 (2020), 1–19; David Nirenberg, "Was There Race Before Modernity? The Example of 'Jewish' Blood in Late Medieval Spain," in *The Origins of Racism in the West*, ed. Miriam Eliav-Feldon, Benjamin Isaac, and Joseph Ziegler (Cambridge: Cambridge University Press, 2009), 232–64; Asa S.

Mittman, "Are the 'Monstrous Races' Races?," *Postmedieval: A Journal of Medieval Cultural Studies* 6, no. 1 (2015): 36–51; Thomas Hahn, "The Difference the Middle Ages Makes: Color and Race Before the Modern World," *Journal of Medieval and Early Modern Studies* 31, no. 1 (Winter 2001): 1–37; Benjamin Isaac, *The Invention of Racism in Classical Antiquity* (Princeton, NJ: Princeton University Press, 2006); Geraldine Heng, *The Invention of Race in the European Middle Ages* (Cambridge: Cambridge University Press, 2018); Scott Pearce, "The Inquisitor and the Moseret: The Invention of Race in the European Middle Ages and the New English Colonialism in Jewish Historiography," *Semantic Scholar*, August 25, 2020, https://www.semanticscholar.org/paper/The-Inquisitor-and-the-Moseret%3A-The-Invention-of-in-Pearce/5c334fa4e22f1c6f8f79d70ebcc9af4b9a60e56b. See also William Chester Jordan, "Why 'Race'?," *Journal of Medieval and Early Modern Studies* 31, no. 1 (Winter 2001): 165–73. For useful and interesting essays by Jordan and others on the ways in which differences now construed as racial were constructed before the arrival of the concept of race, see Thomas Hahn, ed., *A Cultural History of Race in the Middle Ages*, vol. 2 (New York: Bloomsbury, 2023).

4. Hannaford, *Race*, 51, 8–9, 6.

5. "AABA Statement on Race & Racism," American Association of Biological Anthropologists, March 27, 2019, https://bioanth.org/about/position-statements/aapa-statement-race-and-racism-2019/?ref=quillette.com; Joseph L. Graves Jr. and Alan H. Goodman, *Racism Not Race: Answers to the Most Frequently Asked Questions* (New York: Columbia University Press, 2022), 9. The science of the concept of race is dealt with briefly in Graves and Goodman, *Racism Not Race*, 21–41; H. F. Augstein, introduction to *Race: The Origins of an Idea, 1760–1850*, ed. H. F. Augstein (Bristol: Thoemmes Press, 1996), ix–x; "Jena Declaration," Friedrich-Schiller-Universität Jena, https://www.uni-jena.de/en/190910-jenaererklaerung-en. In a similar vein, Robert Bernasconi describes his collection of recent critical essays as a series of attempts to address "Continental Philosophy's racism," beginning with "Kant's racism," Hume's racism, and the racism implicit in "cosmopolitanism" generally. Robert Bernasconi, introduction to *Race and Racism in Continental Philosophy*, ed. Robert Bernasconi and Sybol Cook (Bloomington: Indiana University Press, 2003), 1–7, 2. In his introduction to *Race and the Enlightenment: A Reader* (Malden, MA: Blackwell, 1997), Emmanuel Chukwudi Eze says that his first intention was to title his book "Racist Enlightenment" (1). The case for the racism of racial theory is made directly in C. Loring Brace, *"Race" Is a Four-Letter Word: The Genesis of the Concept* (New York: Oxford University Press, 2005); Robert Wald Sussman, *The Myth of Race: The Troubling Persistence of an Unscientific Idea* (Cambridge, MA: Harvard University Press, 2014). See also George Frederickson, *Racism: A Short History* (Princeton, NJ: Princeton University Press, 2015). It is worth noting that the English term racism in its modern sense did not appear with any frequency until the late 1930s, with its use accelerated by Ruth Benedict's widely disseminated 1940 book *Race: Science and Politics* (1940; repr., Athens: University of Georgia Press, 2019), which begins with a brief discussion of "Racism: The 'Ism' of the Modern World," 3–8. In his 1942 book *Man's Most Dangerous Myth: The Fallacy of Race* (New York: Harper, 1942), Ashley Montagu produced a monumental three-page footnote (87–89) listing publications in which a connection was drawn between slavery and the premise of Negro inferiority. The

connection between the study of race and racist attitudes is disputed by Nancy Stepan, who says in her history of British racial thought that "though many of the scientists who studied race were guilty of bias in the collection and interpretation of their data, of failing to consider contrary evidence, and of making hasty or facile generalisations, few of them knowingly broke the accepted canons of scientific procedure of their day. Most of them were not consciously racist. Many were instead people of humane outlook, opponents of slavery, decent individuals who would have been shocked by any charge that they were racists." Stepan, *The Idea of Race in Science*, xiv.

6. Throughout this book, terms used for racial groups will generally reflect the conventions of the time and place being described or the sources between referred to. I will capitalize white and black only when quoting sources in which they are capitalized, including translations. The use of the capital letter for black is a recent innovation with a sinister history, being the preferred usage of, for example, John C. Calhoun and others who wished to establish a categorical distinction between slaves and white people. The customary usage of Frederick Douglass, W.E.B. Du Bois, comte de Gobineau, Anténor Firmin (although not his translator), and more recently the historian John Hope Franklin was the less restrictive and exclusive lowercase.

7. Charles Darwin, *The Descent of Man, and Selection in Relation to Sex*, 2 vols. (London: John Murray, 1871), 174. For an authoritative discussion of Darwin's relation to French thinking on race in particular, see Bronwen Douglas, "Darwin and the French: The Species Question and 'Man' in Oceania," *Studies in History and Philosophy of Science* 91 (February 2022): 168–80. On the relationship between Darwinian biology and theories of racial extinction, see Patrick Brantlinger, *Dark Vanishings: Discourse on the Extinction of Primitive Races, 1800–1930* (Ithaca, NY: Cornell University Press, 2003).

8. See Thomas F. Gossett, *Race: The History of an Idea in America* (1963; repr., New York: Oxford University Press, 1997), 66. As Gossett wrote in that book, "The importance of Negro slavery in generating race theories in [the United States] can hardly be overestimated" (29). On this point, see also William Stanton, *The Leopard's Spots: Scientific Attitudes Towards Race in America, 1815–59* (Chicago: University of Chicago Press, 1960), 194.

9. Emmanuelle Sibeud, "A Useless Colonial Science? Practicing Anthropology in the French Colonial Empire, Circa 1880–1960," *Current Anthropology* 53 (2012): 55, https://www.journals.uchicago.edu/doi/10.1086/662682.

10. Kidd, *The Forging of Races*, 54.

11. In the first nine editions of *Systema Naturae*, Linnaeus had listed *Homo variat* (literally, "Man varies") under the species Homo, but in the tenth edition this phrase was dropped. Turton translated the thirteenth edition, which had been expanded by J. F. Gmelin.

12. Thomas Jefferson, *Notes on the State of Virginia*, query XIV, Library of Congress, https://tile.loc.gov/storage-services/service/gdc/lhbcb/04902/04902.pdf.

13. For a useful overview of travel narratives, see Jerry Bentley, "Travel Narratives," World History Sources, https://chnm.gmu.edu/worldhistorysources/unpacking/travelmain.html.

14. J. C. Nott and Geo. R. Gliddon, *Types of Mankind: or, Ethnological Researches, Based upon the Ancient Monuments, Paintings, Sculptures, and Crania of Races, and*

upon Their Natural, Geographical, Philological and Biblical History, 6th ed. (Philadelphia: Lippincott, Grambo & Co., 1854). See also Herbert H. Odom, "Generalizations on Race in Nineteenth-Century Physical Anthropology," *Isis* 58, no. 1 (Spring 1967): 4–18.

15. John Beddoe, *The Races of Britain: A Contribution to the Anthropology of West Europe* (London: Trübner, 1885), a county-by-county record of Britain's physical anthropology undertaken before the railroad could confuse older settled populations. William Z. Ripley, *The Races of Europe: A Sociological Study* (New York: Appleton and Co., 1899), 395–96. See also Joseph Jacobs, "Are Jews Jews?," *Popular Science Monthly* 55 (August 1899): 502–11, https://en.wikisource.org/wiki/Popular_Science_Monthly/Volume_55/August_1899/Are_Jews_Jews%3F.

16. Gossett, *Race*, 9.

17. Joseph Conrad, *Heart of Darkness*, ed. Paul B. Armstrong (New York: W. W. Norton & Company, 2006), 11–12.

18. As Winthrop D. Jordan describes the postclassical applications of the chain of being, the concept "served to dramatize the Christian view of man as a creature with a divine soul; it served to formulate men's vague sense of the beast within themselves and their capacity for rising above bestiality; it served to satisfy the eighteenth century's ravenous appetite for hierarchical principles in the face of social upheaval; and it served as a powerful means of organizing the facts of the natural world." Winthrop D. Jordan, *White Over Black: American Attitudes toward the Negro, 1550–1812* (Baltimore: Penguin Books, 1968), 220; see 216–34.

19. Hannah Arendt, *The Origins of Totalitarianism* (1948; repr., Orlando: Harvest Book, 1976), 157. Race was also, in Arendt's analysis, a deep threat to the idea of international law, the "sole regulating principle" of which was an inclusive understanding of humanity.

20. George-Louis Leclerc (Comte de Buffon), *The Natural History of Man*, vol. 3, *Natural History, General and Particular* (1749; repr., Edinburgh: William Creech, 1780), 404–5. Buffon is sometimes linked with Linnaeus as the cofounder of anthropology. But in substantive matters, the two had little in common. Buffon rejected the theory associated with Linnaeus that species were constant and unchangeable, and also the idea that nature was organized as a system of categories. He posited an original homogeneous species of humans that subsequently became differentiated in diverse climates, and devoted much of his effort to finding explanations for the biological processes of differentiation.

21. Samuel George Morton, *Crania Americana, or, A Comparative View of the Skulls of Various Aboriginal Nations of North and South America* (Philadelphia: J. Dobson, 1839), 5.

22. Nott and Gliddon, *Types of Mankind*, 82–83, 61.

23. James Cowles Prichard, *Researches into the Physical History of Mankind*, 2nd ed. (London: John and Arthur Arch, 1826), 1:90. As Bronwen Douglas notes, "Before the publication of Darwin's *On the Origin of Species* (1859), the concept of species was given diverse, often ambiguous meanings depending on a shifting constellation of relative emphases: on reproductive or morphological criteria; history or taxonomy; environment or heredity; hybridization or racial purity; and transmutation or fixity." Douglas, "Climate to Crania," 58. On the loosening and increasing inconsistency of

the scientific understanding of species, see also Bronwen Douglas, "Darwin and the French," 171–74. Nor was variety more clearly defined than species or race. In *The Origin of Species*, Darwin commented that "I look at the term species, as one arbitrarily given for the sake of convenience to a set of individuals closely resembling each other, and that it does not essentially differ from the term variety, which is given to less distinct and more fluctuating forms. The term variety, again, in comparison with mere individual differences, is also applied arbitrarily, and for mere convenience sake." Like Prichard, Darwin noted the challenges arising from resemblance, with some "varieties mocking, as it were, some of the other species of the same genus." Charles Darwin, *On the Origin of Species by Means of Natural Selection* (London: John Murray, 1859), 52, 62. On the confusion of terminology at this time, see John S. Haller Jr., "The Species Problem: Nineteenth-Century Concepts of Racial Inferiority in the Origin of Man Controversy," *American Anthropologist* 72, no. 6 (December 1970): 1319–29. The century and a half since Darwin has not seen any improvement. The lengthy Wikipedia entry on species is almost entirely given over to objections or alternatives to the concept.

24. Prichard, *Researches into the Physical History of Mankind*, 91, 92.

25. See, for example, John Mitchell, "An Essay upon the Causes of the Different Colours of People in Different Climates," in *Philosophical Transactions* (1744), 43:102–50; Petrus Camper, "On the Origin and Color of Blacks" (1772), trans. Miriam Claude Meijer, "Footnotes to the History of Anthropology: Petrus Camper on the Origin and Color of Blacks," *History of Anthropology Newsletter* 24, no. 2 (1997): article 3, 4–9; Sir William Jones, "Discourse the Ninth: The Origin and Families of Nations" (1792), in *The Works of Sir William Jones* (London: John Shore, 1807), 9:185–204; Morton, *Crania Americana.*

26. The term polygenism did not come into circulation until about 1860, but the theory of plural origins, also called plurigenesis, was established long before that. As Blumenbach noted, "Ill-feeling, negligence, and the love of novelty have induced persons to take up the [polygenist] opinion. The idea of the plurality of human species has found particular favour with those who made it their business to throw doubt on the accuracy of Scripture." Johann Friedrich Blumenbach, *On the Natural Variety of Mankind* (1755), in *The Anthropological Treatises of Johann Friedrich Blumenbach*, ed. and trans. Thomas Bendyshe (1790; repr., London: Longman, Green, Longman, Roberts, and Green, 1865), 65–144.

27. Nott and Gliddon, *Types of Mankind*, 74.

28. Ruth Benedict and Gene Weltfish, "The Races of Mankind" (1943), in *Race: Science and Politics*, by Ruth Benedict (1940; repr., Athens: University of Georgia Press, 2019), 171. On the role played by religion in the scientific study of race, see Terrence Keel, *Divine Variations: How Christian Thought Became Racial Science* (Stanford, CA: Stanford University Press, 2018).

29. Arendt, *Origins*, 159.

30. Jules Michelet, *History of France*, trans. G. H. Smith (1831; repr., New York: D. Appleton and Company, 1882), 1:67; Ernest Renan, "Islam and Science" (1883), in *What Is a Nation? And Other Political Writings*, ed. and trans. M.F.N. Giglioli (New York: Columbia University Press, 2018), 265. Renan had no difficulty characterizing and ranking races, and defending colonial domination on that basis. See Ernest

Renan, "Intellectual and Moral Reform of France" (1871), in *What Is a Nation? And Other Political Writings*, ed. and trans. M.F.N. Giglioli (New York: Columbia University Press, 2018), 182–246. On French racial science, see Claude Blanckaert, "On the Origins of French Ethnology: William Edwards and the Doctrine of Race," in *Bones, Bodies, Behavior: Essays on Biological Anthropology*, ed. George W. Stocking Jr. (Madison: University of Wisconsin Press, 1988), 18–55; Martin S. Staum, *Labeling People: French Scholars on Society, Race, and Empire 1815–1848* (Montreal: McGill-Queen's University Press, 2003); Ian B. Stewart, "William Frédéric Edwards and the Study of Human Races in France, from the Restoration to the July Monarchy," *History of Science: An Annual Review of Literature, Research and Teaching* 58, no. 3 (December 2019), https://journals.sagepub.com/doi/10.1177/0073275319889832; Alice Conklin, *In the Museum of Man: Race, Anthropology, and Empire in France, 1850–1950* (Ithaca, NY: Cornell University Press, 2013). For a discussion of nationalism and "racism" with a particular focus on France and Germany, see Ruth Benedict, *Race: Science and Politics* (1940; repr., Athens: University of Georgia Press, 2019), esp. 127–39.

31. The inclusive conception of French citizenry is no longer uncontested. In recent years a different understanding of French national identity has arisen, driven by a "replacement theory" of immigration in which differences that the political Left construes as a proof of the ability of the nation to absorb variations are represented by the Right as unassimilable threats to the identity of the nation. See Emmanuelle Saada, "*Un racisme de l'expansion. Les discriminations raciales au regard des situations coloniales*," in *De la question sociale à la question raciale?*, ed. Didier Fassin and Eric Fassin (Paris: La Découverte, 2006), 55–71. Jean Raspail's novel *The Camp of the Saints* has been widely seen as an influential text in the development of replacement theory. Jean Raspail, *The Camp of the Saints*, trans. Norman Shapiro (1973; repr., New York: Charles Scribner's Sons, 1975). On Martinique, see Franz Fanon, *Black Skins, White Masks*, trans. Haakon Chevalier (New York: Grove Books, 2008), originally published as *Peau noire, masques blancs* (1952).

32. Arthur, Comte de Gobineau, *Essai sur l'inégalité des races humaines* [Essay on the inequality of human races], trans. Adrian Collins (1853–55; repr., London: Heinemann, 1915).

33. Arendt, *Origins*, 165.

34. Arendt, *Origins*, 174.

35. Renan's thinking is controversial in part because it is inconsistent. Known as a great advocate of tolerance, he is also associated with antisemitism and linked with Gobineau. His *La Réforme Intellectuelle et Morale* includes descriptions of the Chinese as a "race of workers," Negroes as "tillers of the soil," and Europeans as "a race of masters and soldiers." Ernest Renan, *La Réforme Intellectuelle et Morale*, 4th ed. (1871; repr., Paris: Michel Lévy Frères, 1875), 94. On the complex relations between Renan and his interpreters, see Dora Bierer, "Renan and His Interpreters: A Study in French Intellectual Warfare," *Journal of Modern History* 25, no. 4 (1953): 375–89.

36. For essays on the German role in the development of racial theory prior to 1850, see Sara Eigen and Mark Larrimore, eds., *The German Invention of Race* (Albany: SUNY Press, 2006).

37. Daniel Defoe, *The True-Born Englishman: A Satire* (London, 1701). For a version with the spelling and punctuation modernized; see https://www.gutenberg.org/files/30159/30159-h/30159-h.htm. On the shaping of English national identity, see Linda Colley, *Britons: Forging the Nation, 1707–1837* (London: Pimlico, 1994).

38. See "Anglo-Saxonism, Polygenism, and Physical Anthropology," in *Victorian Anthropology*, by George W. Stocking Jr. (New York: Free Press, 1987), 62–69.

39. Benjamin Disraeli, *Tancred: or, The New Crusade*, vol. 1 (London: Henry Colburn, 1847), 303; Robert Knox, *The Races of Men* (Philadelphia: Lea & Blanchard, 1850), 7.

40. Thomas Huxley, "The Forefathers and Forerunners of the English People," *Pall Mall Gazette*, January 10, 1870, 8–9, reprinted in Michael D. Biddiss, *Images of Race* (New York: Holmes & Meier, 1979), 168.

41. Frederickson, *Racism*, 75. See also Nicholas Hudson, "From 'Nation' to 'Race': The Origin of Racial Classification in Eighteenth-Century Thought," *Eighteenth-Century Studies* 29, no. 3 (Spring 1996): 247–64. For one, among many, discussions of race and nation in American political thinking, see Gary Gerstle, *American Crucible: Race and Nation in the Twentieth Century* (Princeton, NJ: Princeton University Press, 2017). See also Geoffrey Galt Harpham, *Citizenship on Catfish Row: Race and Nation in American Popular Entertainment* (Columbia: University of South Carolina Press, 2023). Richard McMahon has argued that in more recent years, race has been deployed in "discourses of northern superiority and eastern inferiority that still infuse Euroskepticism and discussion of EU enlargement." Richard McMahon, *The Races of Europe: Construction of National Identities in the Social Sciences, 1839–1939* (London: Palgrave, 2016), 4. This book is a remarkably rich trove of statistical information, including a database of quantitative bibliographic citation indexes and series of case studies; particularly useful are those concerning Ireland and Poland. See also Richard McMahon, ed., *National Races: Transnational Power Struggles in the Sciences and Politics of Human Diversity, 1840–1945* (Lincoln: University of Nebraska Press, 2019).

42. The first such society of which there is any record is the Société des Observateurs de l'Homme, founded in Paris in 1800.

43. Stocking, *Victorian Anthropology*, 301.

44. Ernst Haeckel, *The History of Creation: or the Development of the Earth and Its Inhabitants by the Action of Natural Causes*, trans. E. Ray Lankester, vol. 2 (1868; repr., London: D. Appleton & Co., 1880); Paul Broca, *On the Phenomena of Hybridity in the Genus Homo* (London: Longman, Green, Longman, Roberts, and Green 1864), originally published as "Des phénomènes d'hybridité dans le genre humain," *Journal de la physiologie de l'homme et des animaux* 3 (1859–60): 392–43; Jean-Louis Armand de Quatrefages de Bréau, "Formation of Races Under the Sole Influence of Conditions of Life and Heredity," in *The Human Species* (1877; repr., London: Kegan Paul & Company, 1879), 244–59; Rudolf Virchow, "Heredity and the Formation of Races," in *This Is Race: An Anthology Selected from the International Literature on the Races of Man*, ed. Earl W. Count (New York: Henry Schuman, 1950), 178–93, originally published as "Rassenbildung und Erblichkeit," in *Festschrift für Adolf Bastian* (Berlin, 1896), 1–43.

45. For a comprehensive argument about the connection between race and modern disciplines including such quantitative disciplines as data science, see Elise K. Burton and Sayori Ghoshal, "The History of Science Through the Prism of Race," Cambridge University Press, May 23, 2024, https://www.cambridge.org/core/journals/bjhs-themes/article/history-of-science-through-the-prism-of-race/D041D3BDF1CAE769C11A2CED4694DD23.

Chapter Two: One or Many?

1. M. Flourens, "Memoir of Blumenbach," in *The Anthropological Treatises of Johann Friedrich Blumenbach*, ed. and trans. Thomas Bendyshe (1790; repr., London: Longman, Green, Longman, Roberts, and Green, 1865), 63.

2. Flourens, "Memoir of Blumenbach," 52.

3. The bust was at the Johann-Friedrich-Blumenbach Institute for Zoology and Anthropology. See "Blumenbach," Vernetzung Göttingen Postkolonial, https://goettingen-postkolonial.de/blumenbach/. In the same year and in furtherance of the same antiracist goals, David Hume's name was removed from Hume Tower at the University of Edinburgh.

4. Flourens, "Memoir of Blumenbach," 50.

5. Flourens, "Memoir of Blumenbach," 50–51.

6. Many of Blumenbach's collected materials are preserved in the Blumenbach collection at the University of Göttingen. See https://blumenbach-online.de/index-englisch.php.

7. Johann Friedrich Blumenbach, *On the Natural Variety of Mankind*, in *The Anthropological Treatises of Johann Friedrich Blumenbach*, ed. and trans. Thomas Bendyshe, 3rd ed (1775; repr., 1795), 145–276. Hereafter cited parenthetically in the text as NV, 1 or NV, 3. See https://archive.org/details/anthropologicalt00blumuoft.

8. This is the judgment of M. F. Ashley Montagu in "The Genetical Theory of Race, and Anthropological Method," *American Anthropologist* 44, no. 3 (July–September 1942): 369–75.

9. Flourens, "Memoir of Blumenbach," 60.

10. Twelve editions of *Systema Naturae* were published in Linnaeus's lifetime, the most important of which was the tenth (1758). Turton's seven-volume English translation of the thirteenth edition, *A General System of Nature, Through the Three Grand Kingdoms of Animals, Vegetables, and Minerals*, was published in London in 1802–6.

11. Linnaeus, letter to Johann Georg Gmelin, February 25, 1747. Gmelin would later become the editor of *Systema Naturae* after Linnaeus' death. Quoted in Nancy Stepan, *The Idea of Race in Science: Great Britain, 1800–1960* (Houndmills, UK: Macmillan, 1982), 7.

12. Flourens ("Memoir of Blumenbach," 53) writes that "everything belonging to method was neglected by Blumenbach; he confined himself to following Linnaeus; he adopted from him almost all his *divisions* with whatever advantages they had, and also with all their defects, their narrowness of study, and their caprice."

13. J. F. Blumenbach, *Abbildungen naturhistorischer Gegenstände*, 1st ed., nos. 1–10 (Göttingen: Johann Christian Dieterich, 1796), 25–27. On this text, see John S. Michael, "*Porträts von interessanten Personen*: A New Look at J. F. Blumenbach's

Typological Labels and the Exemplars He Discussed in His Anthropological Research," *Annals of the History of Philosophy and Biology* (2020): 65–101, 71–72, https://univerlag.uni-goettingen.de/bitstream/handle/3/eissn-2512-5923_annals25/Annals25.pdf?sequence=1&amp. See also Johann Friedrich Blumenbach, "Observations on Bodily Conformation and Mental Capacities of the Negroes," *Philosophical Magazine* 3 (1799): 144–47, https://www.tandfonline.com/doi/abs/10.1080/14786449908676972.

14. In the 1795 edition of *Natural Variety*, Blumenbach introduced the term *nigrita*, translated as Negro and meaning black person. According to Wolfgang Böker, to whom I owe this information, the term was not used in the first or second editions.

15. Johann Friedrich Blumenbach, *Contributions to Natural History*, 2nd ed. (1790; repr., 1806), in *The Anthropological Treatises of Johann Friedrich Blumenbach*, ed. and trans. Thomas Bendyshe (1790; repr., London: Longman, Green, Longman, Roberts, and Green, 1865), 312. Blumenbach had made similar statements as early as 1787. For "*our black brothers*," see Flourens, "Memoir of Blumenbach," 60.

16. For a rare comparable statement from a contemporary of Blumenbach, see Henri Grégoire, *De la litterature des Nègres: ou recherches sur leurs facultés intellectuelles, leurs qualités morales et leur littérature; suivies de notices sur la vie et les ouvrages de Nègres qui se sont distingués dans les sciences, les lettres et les arts* [Of the literature of the Negroes: or research on their intellectual faculties, their moral qualities and their literature; followed by notes on the life and works of Negroes who distinguished themselves in the sciences, letters and the arts] (Paris: Chez Maradan, 1808).

17. In the 1795 edition, Blumenbach revisited this theme with a certain gusto, rehearsing "horrid stories . . . of the union of men with brutes, when either men have had to do with the females of beasts (whether carried away by unbridled lust, or from some mad idea of continence, or because they expected some medicinal aid from this sort of crime), or when we are told that women have been made use of by male brutes (whether that has happened through any violent rape, or because women have solicited them in the madness of lust, or have prostituted themselves from religious superstition)" (NV, 3:201). There is, he hastens to add, no evidence that any such connection has been fruitful.

18. His example is the conceit of a "regular gradation" of forms in nature—the same concept used by Charles White in a 1795 address given to the Literary and Philosophical Society of Manchester. The lecture was published four years later with the title *An Account of the Regular Gradation in Man*.

19. The climate theory of racial difference was first proposed by Gottfried Wilhelm Leibniz in 1697. Acknowledging superficial differences between peoples, Leibniz insisted that there "is no reason why all men who inhabit the earth should not be of the same race [*d'une même race*], which has been altered by different climates." Gottfried Wilhelm von Leibniz, "Lettre de Mr. Leibniz à Mr. Sparvenfeld" (1697), in *Otium Hanoveranum sive Miscellanes ex ore* . . . , ed. Joachim Friedrich Feller (Leipzig: Johann Christiani Martini, 1718), 38. See https://digitale.bibliothek.uni-halle.de/vd18/content/pageview/5315627.

20. Blumenbach, *Contributions to Natural History*, 294.

21. See Hippocrates, "Airs, Waters, Places," in *Hippocratic Writings*, ed. G.E.R. Lloyd, trans. J. Chadwick and W. N. Mann (London: Penguin Press, 1950), 148–69. For an account of the role played in racial science by the nonscientific discourse of religion, see Terrence Keel, *Divine Variations: How Christian Thought Became Racial Science* (Stanford, CA: Stanford University Press, 2018), esp. 23–54. Keel argues that "scientific" ideas about race remained tacitly invested in Christian ideas about creation, ancestry, and universalism.

22. Johann Friedrich Blumenbach, *A Manual of the Elements of Natural History* [Handbuch der Naturgeschichte], trans. R. T. Gore from 10th German ed. (1779; repr., London: W. Simpkin and R. Marshall, 1825), 12.

23. See Petrus Camper, *Dissertation sur les variétés naturelles qui caractérisent la physionomie des hommes des divers climats et des différens âges*, trans. H. J. Jansen from Dutch to French (Paris: H. J. Jansen, 1791). For Camper's views, see Petrus Camper, "On the Origin and Color of Blacks" (1772), trans. Miriam Claude Meijer, in "Footnotes to the History of Anthropology: Petrus Camper on the Origin and Color of Blacks," *History of Anthropology Newsletter* 24, no. 2 (1997): article 3, 4–9, https://petruscamper.com/han.htm. In this text, Camper argues that "all of us are black, only more or less."

24. After the publication in 1859 of Darwin's *Origin of Species*, such serial presentations would be used and interpreted by some as illustrations of evolutionary progress.

25. As historian Nancy Stepan notes with considerable understatement, "It is a curious fact of history that differences in skull shapes between peoples that today seem relatively inconsequential, and certainly insufficient for the purposes of racial categorization, were by the end of the eighteenth and throughout the nineteenth centuries taken to be of prime significance for distinguishing the geographical and local races of man." Nancy Stepan, *The Idea of Race in Science: Great Britain, 1800–1960* (Houndmills, UK: Macmillan, 1982), 33.

26. K.F.H. Marx, "Life of Blumenbach," in *The Anthropological Treatises of Johann Friedrich Blumenbach*, ed. and trans. Thomas Bendyshe (1790; repr., London: Longman, Green, Longman, Roberts, and Green, 1865),8.

27. This, at least, is the theory propounded by Nell Irvin Painter in *The History of White People* (New York: W. W. Norton & Company, 2010), 82.

28. At the conclusion of his investigation of the roles played by skulls in Friedrich von Schiller and Blumenbach, Robert J. Richards writes, "I have seen the Georgian female's skull in the Blumenbach collection at the Georg-August-Universität in Göttingen. It is quite beautiful." Robert J. Richards, "The Beautiful Skulls of Schiller and the Georgian Girl: Quantitative and Aesthetic Scaling of the Races, 1770–1850," in *Johann Friedrich Blumenbach: Race and Natural History, 1750–1850*, ed. Nicolaas Rupke and Gerhard Lauer (London: Routledge, 2019), 172, https://home.uchicago.edu/~rjr6/articles/Skulls%20for%20Blumenbach%20final%20Chapter%209.pdf.

29. Blumenbach wrote in *Natural Variety* that "the more my daily experience and, as it were, my familiarity with my collection of skulls of different nations increases, so much the more impossible do I find it to reduce these racial varieties . . . to the measurements and angles of any single scale" (NV, 3:236).

30. Blumenbach, whose list of close friends included Jewish physician K.F.H. Marx (whose "Life of Blumenbach" was included in the Bendyshe edition of Blumenbach's works), did not figure in any official Nazi publication. See Nicolaas Rupke and Gerhard Lauer, "Introduction: A Brief History of Blumenbach Representation," in *Johann Friedrich Blumenbach: Race and Natural History, 1750–1850*, ed. Nicolaas Rupke and Gerhard Lauer (London: Routledge, 2019), 9.

31. Ernie Hamm, "Christian Thought, Race, Blumenbach, and Historicizing," *Zygon: Journal of Religion and Science* 54, no. 1 (2019): 237–45, https://doi.org/10.1111/zygo.12495 (emphasis added).

32. George M. Frederickson, *Racism: A Short History* (Princeton, NJ: Princeton University Press, 2002), 57 (emphasis added).

33. Stephen Jay Gould, *The Mismeasure of Man* (New York: W. W. Norton & Company, 1996), 48, 410, 412 (emphasis added).

34. Painter, *The History of White People*, 81.

35. Jacob Bronowski, *The Ascent of Man* (London: McDonald Future, 1973), 367 (emphasis added).

36. Bruce Baum, *The Rise and Fall of the Caucasian Race: A Political History of Racial Identity* (New York: NYU Press, 2006), 89 (emphasis added).

37. Emmanuel Chukwudi Eze, ed., *Race and the Enlightenment: A Reader* (Cambridge, MA: Blackwell, 1997), 79 (emphasis added).

38. See Basisgruppe Umweltwissenschaften 2020, "Stellungnahme der Basisgruppe Umweltwissenschaften zu Sturz und Entfernung der Büsten von Haeckel und Blumenbach im Zoologischen Institut," Facebook post, June 30, 2020, http://www.facebook.com/103900431295642/posts/-stellungnahme- der-basisgruppe-umweltwissenschaften-zu-sturz-und-entfernung-der-/152316819787336/ (emphasis added).

39. Søren Kierkegaard, *The Concept of Dread*, trans. Walter Lowrie (Princeton, NJ: Princeton University Press, 1973), also translated in less resonant terms as *The Concept of Anxiety*. See also Søren Kierkegaard, *Fear and Trembling*, trans. Walter Lowrie (Princeton, NJ: Princeton University Press, 1974). On the fall of man, see Dietrich Bonhoeffer, *Creation and Fall* and *Temptation* (New York: Macmillan, 1966).

40. Kierkegaard, *Concept of Dread*, 41.

41. In a representative statement on this topic, Meiners commented that Caucasians had "a more splendid disposition of the mind and heart" (*durch herrlichere Anlagen des Geistes und Herzens*) than other races. Christoph Meiners, *Grundriß der Geschichte der Menschheit* [Outline of the history of mankind] (Lemgo, Germany: Verlag der Meyerschen Buchhandlung, 1793), 75.

42. Christoph Meiners, *Betrachtungen über die Fruchtbarkeit* (Lübeck, Germany: 1795–96), quoted and translated in Britta Rupp-Eisenreich, "Christoph Meiners' 'New Science' (1747–1810)," in *The Invention of Race: Scientific and Popular Representations*, ed. Nicolas Bancel, Thomas David, and Dominic Thomas (London: Routledge, 2014), 74.

43. Georg Johann Adam Forster described Meiners as a man "smitten with his mania-of-knowing-better." Georg Johann Adam Forster, "Meiners anthropologische Abhandlungen," *Göttingisches historisches Magazin*, 4–7, translated and quoted in Britta Rupp-Eisenreich, "Christoph Meiners' 'New Science' (1747–1810),"

in *The Invention of Race: Scientific and Popular Representations*, ed. Nicolas Bancel, Thomas David, and Dominic Thomas (London: Routledge, 2014), 77.

44. Michael C. Carhart argues on the basis of a text published posthumously that Meiners, responding to French theory and reports from South Sea explorers—reports he systematically and willfully misinterpreted—eventually abandoned his initial theory of a dual origin for humanity, and embraced a theory that the different races emerged indigenously at different times and places. See Michael C. Carhart, "Polynesia and Polygenism: The Scientific Use of Travel Literature in the Early Nineteenth Century," *History of the Human Sciences* 22, no. 2 (2009): 58–86, https://philpapers.org/rec/CARPAP-9.

45. Christoph Meiners, *Grundriß der Geschichte der Menschheit* [Outline of the history of mankind], quoted and translated in Britta Rupp-Eisenreich, "Christoph Meiners' 'New Science' (1747–1810)," in *The Invention of Race: Scientific and Popular Representations*, ed. Nicolas Bancel, Thomas David, and Dominic Thomas (London: Routledge, 2014), 71. For a consideration of the controversy with Blumenbach, see Rupp-Eisenreich, "Christoph Meiners," 78–81. For a discussion of Meiners's views on race, see Johan van der Zande, "Christoph Meiners on *Lebenswissenschaft*," in *Christoph Meiners 1747–1810*, ed. Stefan Klingner and Gideon Stiening (Berlin: De Gruyter, 2023), 242–44.

46. Christoph Meiners, "Of the Varieties and Deviate Forms of Negroes," trans. Jon M. Mikkelson, in *Kant and the Concept of Race: Late Eighteenth-Century Writings*, ed. Jon M. Mikkelson (Albany: SUNY Press, 2013), 198–201, originally published as "Von den Varietäten und Abarten der Neger" (1790). See also Christoph Meiners, *Ueber die Natur der afrikanischen Neger und die davon abhangende Befreyung, oder Einschränkung der Schwarzen* [On the nature of the African Negroes and the liberation or restriction of blacks] (1790; repr., Wehrhahn Verlag, 2000); Christoph Meiners, "Ueber die Rechtmäßigkeit des Neger-Handels" [On the lawfulness of the Negro trade], *Göttingisches historisches Magazin* 2, no. 3 (1788): 398–416; Christoph Meiners, "Forgesetzte Betrachtungen über den Sclavenhandel, und die Freylassung der Neger" [Other considerations on the slave trade and the liberation of Negroes], *Neues Göttingisches historisches Magazin* 2 (1793): 1–58.

47. Meiners, "Of the Varieties," 203.

48. John S. Michael, "The Race Supremacist Anthropology of Christoph Meiners, Its Origins and Reception," unpublished manuscript, n.d., 4. Morgan Golf-French has written of Meiners, "For the most part, however, accounts of the period have simply avoided or skirted around the darker corners of the German Enlightenment." Morgan Golf-French, "Beyond Heroes and Villains: Reassessing Racism in the German Enlightenment," Hypotheses, https://ghil.hypotheses.org/339. For a comprehensive approach to Meiners, see Stefan Klingner and Gideon Stiening, eds., *Christoph Meiners 1747–1810* (Berlin: De Gruyter, 2023). In addition to fifteen essays on aspects of Meiners's thought, this volume contains a comprehensive bibliography. Even so, the relation to Blumenbach is examined in only one essay, and that briefly.

49. Rupp-Eisenreich, "Christoph Meiners," 71.

50. Julien-Joseph Virey, *Natural History of the Negro Race*, "Extracted from the French," trans. J. H. Guenebault (Charleston, SC: D. J. Dowling, 1837), 19, originally published as *Histoire Naturelle du genre humaine* (1800). "They appear," Virey

says of Negroes, "to be altogether '*corporeal*,' having scarcely a notion of a superior being, and no idea embodied with any but carnal pleasures. There is but one step between their intellect and that of the orang-outang; in short, they lead an animal life" (Virey, *Natural History of the Negro Race*, 11). See also https://archive.org/details/naturalhistoryn00soemgoog/page/n126/mode/2up?q=VIREY.

51. For an equivocal portrait of Meiners, see Philipp Albert Stapfer, "The Life of Christoph Meiners," trans. John S. Michael, in "The Race Supremacist Anthropology of Christoph Meiners, Its Origins and Reception," by John S. Michael, unpublished manuscript, September 21, 2021, 70–109. Stapfer ("Life of Christoph Meiners," 90) says that Meiners's writings on the inferiority of the Negro race were "cited in debates in the British Parliament by the defenders of that infamous trafficking which has for so long brought the disgrace of the peoples of Europe and the shame of the Christian nations." Stapfer does note that in his final years, Meiners was tasked by the imperial Russian government with recruiting German professors to teach in Russian universities, thus extending the reach of German scholarship and serving German interests.

52. Meiners is mentioned in a footnote by Gobineau in his *Essai sur l'inégalité des races humaines* (Paris, 1851–53), book 1, chap. 10, 121, https://nacaomestica.org/blog4/wp-content/uploads/2017/02/essai_inegalite_races_1.pdf.

53. Egon von Eickstedt, cited in Rupp-Eisenreich, "Christoph Meiners," 69.

54. In *The History of White People*, Painter points to Blumenbach's emphasis on beauty and repeated references to Tacitus as evidence that Blumenbach was influenced by the "fringy" Meiners rather than the other way round. Painter, *History of White People*, 87.

55. A more compelling advocate for the word race was Kant, who legitimated the term by using it along with several others in the course of his many lectures on anthropology. See Immanuel Kant, "Of the Different Human Races" (1777), in *Kant and the Concept of Race: Late Eighteenth Century Writing*, ed. and trans. Jon M. Mikkelson (Albany: SUNY Press, 2013), 55–72. See also Immanuel Kant, "On National Characters in so far as They Rest upon the Different Feeling of the Sublime and the Beautiful" (1764), in *Observations on the Feeling of the Beautiful and Sublime*, ed. Patrick Frierson and Paul Guyer, trans. Paul Guyer (Cambridge: Cambridge University Press, 2011), 50–62. For an efficient account of the Kant-Blumenbach relationship, see Mario Marino, "At the Roots of Racial Classification: Theory and Iconography in the Work and Legacy of Johann Friedrich Blumenbach," in *Bérose-Encyclopédie internationale des histoires de l'anthropologie* (Paris, 2022), https://www.berose.fr/article2546.html?lang=fr. This relationship is also the subject of a careful argument by Robert Bernasconi, who shows that in the course of their exchanges, both Kant and Blumenbach modified their views, with Blumenbach coming to accept the term race, preferred by Kant, with its implication of permanence or fixity, as a way of giving some order to the processes by which hybridity could produce a modification of the *Bildungstrieb*. He did so even while continuing to doubt Kant's emphasis on skin color as a racial marker. Robert Bernasconi, "Kant and Blumenbach's Polyps: A Neglected Chapter in the History of the Concept of Race," in *The German Invention of Race*, ed. Sara Eigen and Mark Larrimore (Albany: SUNY Press, 2006), 73–90. For the argument that Blumenbach's wary ambivalence toward the race concept was better expressed in a predominantly pictorial atlas he produced than in any of his

prose, see Sara Eigen, "Self, Race, and Species: Johann Friedrich Blumenbach's Atlas Experiment," *German Quarterly* 78, no. 3 (2005): 277–98.

56. Johann Friedrich Blumenbach, "Observations on Some Egyptian Mummies Opened in London," *Philosophical Transactions of the Royal Society of London* 84 (1794): 193.

57. Blumenbach, *Contributions to Natural History*, 290–91. In this text, too, Blumenbach uses the term *Rasse* on several occasions. Wolfgang Böker has drawn my attention to two studies of Blumenbach's use of *Rasse*. Renato G. Mazzolini, "Quale lingua per la scienza? Traduzioni di testi scientifici di italiani e tedeschi nel secondo Settecento," in *Traduzione e transfert nel XVIII secolo. Tra Francia, Italia e Germani*, ed. Giulia Cantarutti and Stefano Ferrari (Milan: FrancoAngeli, 2013), 67–93; Renato G. Mazzolini, "*Race* versus *varietas*," in *Le parole del pensiero. Studi offerti a Nestore Pirillo*, ed. Fabrizio Meroi (Pisa: Edizioni ETS, 2013), 91–10.

58. James Hunt, *On the Negro's Place in Nature* (London: Trübner, 1863), 6. For a survey of the ways Blumenbach has been "translated" into Meiners, see Nicolaas Rupke and Gerhard Lauer, "Introduction: A Brief History of Blumenbach Representation," in *Johann Friedrich Blumenbach: Race and Natural History*, ed. Nicolaas Rupke and Gerhard Lauer (London: Routledge, 2019), 6–9.

59. John S. Michael, "Nuance Lost in Translation," *NTM Zeitschrift für Geschichte der Wissenschaften, Technik und Medizin* 25, no. 3 (2017): 281–309, doi:10.1007/s00048-017-0173-8.

60. Thomas Bendyshe, "Editor's Preface," in *The Anthropological Treatises of Johann Friedrich Blumenbach*, ed. and trans. Thomas Bendyshe (1790; repr., London: Longman, Green, Longman, Roberts, and Green, 1865), vii–xiv, xi, x. The reference is to Isidore Geoffroy Saint-Hilaire, *Classification Anthropologique. Mêm. de la Soc. d'Anthrop. de Paris* (1860), 1:129.

61. Gould's *The Mismeasure of Man* in particular is exposed to criticism for drawing conclusions about Blumenbach from poorly analyzed misinformation. See Thomas Junker, "Blumenbach's Racial Geometry," *Isis* 89, no. 3 (1998): 498–501.

Chapter Three: More or Less Human?

1. Josiah C. Nott, "Statistics of Southern Slave Population with Especial Reference to Life Insurance," *Commercial Review of the South and West* 3 (November 1847): 275–87, https://www.proquest.com/docview/124720799/fulltextPDF/F2312DDDC15E4224PQ/1?accountid=10598. Quotations in the text are taken from this unpaginated text. In a previous issue, Nott had commented on the insurance business in the South but had only considered white people. The journal was also known as *De Bow's Commercial Review*.

2. Josiah C. Nott, "On the Pathology of Yellow Fever," *American Journal of the Medical Sciences* 9 (1845): 277–91. See also Josiah C. Nott, "Yellow Fever Contrasted with Bilious Fever," *New Orleans Medical and Surgical Journal* 4 (1848): 563–604. For a thorough discussion of Nott's contributions to the study of yellow fever, see Reginald Horsman, *Josiah Nott of Mobile: Southerner, Physician, and Racial Theorist* (Baton Rouge: Louisiana State University Press, 1987), 133–49, 150–69.

3. Samuel George Morton, *Crania Americana, or, A Comparative View of the Skulls of Various Aboriginal Nations of North and South America* (Philadelphia: J. Dobson, 1839).

4. Six years after this article appeared, four of Nott's children would die of yellow fever in a single week. Two other sons died fighting for the Confederacy, at Shiloh in 1862 and Chickamauga in 1863, respectively.

5. In a July 15, 1845, letter to Morton, Nott used the phrase "affinity of blood" as a measure of biological distance between races. Morton Papers, quoted in Horsman, *Josiah Nott*, 99.

6. Samuel George Morton, "Hybridity in Animals and Plants, Considered in Reference to the Question of the Unity of the Human Species" (New Haven, CT: B. L. Hamlin, printer to Yale College, 1847), 22. In his effort to naturalize an emotional response, Morton alluded to Blumenbach's assertion that humans had never been in a wild state perhaps because they had, in distant times, been kept as domesticated animals by a now vanished species of superhuman creatures. Johann Friedrich Blumenbach, *Contributions to Natural History* [*Beyträge zur Naturgeschichte*], in *The Anthropological Treatises of Johann Friedrich Blumenbach*, ed. and trans. Thomas Bendyshe (1790; repr., London: Longman, Green, Longman, Roberts, and Green, 1865), 294.

7. Morton, "Hybridity in Animals and Plants," 22.

8. In *Types of Mankind*, published in 1852, Nott modified this view, saying that the Egyptians of "the pyramidal period" were not to be considered "Caucasians" but rather were a mixture of "foreign stocks," chiefly Libyan and Asiatic, on an "aboriginal *African type*"; Negroes were, however, apparently unchanged from their primitive ancestors. Nott and Gliddon, *Types of Mankind*, 229.

9. Philanthropist, "Vital Statistics of Negroes and Mulattoes," *Boston Medical and Surgical Journal* 27, no. 10 (October 12, 1842): 168–170, https://www.nejm.org/doi/full/10.1056/NEJM184210120271003.

10. In considering the statistics of the insurance industry, Nott was ahead of his time. The industry did not begin to rely on quantification until the early twentieth century. See Dan Bouk, *How Our Days Became Numbered: Risk and the Rise of the Statistical Individual* (Chicago: University of Chicago Press, 2015).

11. Josiah C. Nott, "The Mulatto a Hybrid—probable extermination of the two races if the Whites and Blacks are allowed to intermarry," *American Journal of the Medical Sciences* 6, no. 11 (1843): 252–53, https://www.nejm.org/doi/full/10.1056/NEJM184308160290201.

12. Edward Long, *The History of Jamaica or, General Survey of the Antient and Modern State of That Island*, 3 vols. (London: T. Lowndes, 1774), https://repository.library.northeastern.edu/downloads/neu:m0410b06w?datastream_id=content. A section of vol. II, book III, chap. 1, "Negroes," was printed in *The Columbian Magazine* in early 1788 with the title "Observations on the Gradation in the Scale of Being Between the Human and Brute Creation. Including Some Curious Particulars Respecting Negroes."

13. The concept of a ranked or ordered "chain" or "scale" of the world's creatures is found in Aristotle's biology, but had been adapted in various ways over the centuries. It was used explicitly in connection with racial theory in a text Nott may have

known: Charles White, *An Account of the Regular Gradation in Man: And in Different Animals and Vegetables; and from the Former to the Latter* (London: C. Dilly, 1799). While disavowing any approval of slavery, White was one of the first to argue for racial ranking, with Celtic Germans on top and Negroes on the bottom of the scale, just above primates.

14. Long, *History of Jamaica*, 363.

15. Long, *History of Jamaica*, 270, 356.

16. Long, *History of Jamaica*, 371.

17. Thomas Jefferson, *Notes on the State of Virginia*, query XIV, Library of Congress, https://tile.loc.gov/storage-services/service/gdc/lhbcb/04902/04902.pdf.

18. Long, *History of Jamaica*, 328.

19. Long, *History of Jamaica*, 335.

20. Julien-Joseph Virey, *Natural History of the Negro Race Extracted from the French*, trans. J. H. Guenebault (Charleston, SC: D. J. Dowling, 1837), 2–56, originally published as *Histoire Naturelle du genre humaine* (1800). Guenebault's "extraction" included an extended quotation from Jefferson's *Notes on the State of Virginia* that did not make it clear that the words were Jefferson's and not Virey's, and even included references to Virey himself that were obviously not in the original. In his "Dedication" to the members of the Literary and Philosophical Society of Charleston, Guenebault proleptically defended himself against accusations that he was misrepresenting or even traducing the text he was ostensibly translating by suggesting that Virey's text had included some "incorrect expressions" that he had put right.

21. Nott, "The Mulatto a Hybrid," 32.

22. Virey, *Natural History*, 35, 104–5, 3. In a signed entry on "Man" in the *Dictionnaire des sciences naturelle* (Paris, 1816), Virey reported the beliefs that Negroes had copulated with apes, and that Negro blood, like their skin, was black (XV:152), quoted in Léon Poliakov, *The Aryan Myth: A History of Racist and Nationalist Ideas in Europe*, trans. Edmund Howard (New York: Basic Books, 1971), 181. In his edition of Virey, Guenebault included an extract from a 1785 essay by Thomas Soemmering, "On the Comparative Anatomy of the Negro and European," which added a further proof of Negro inferiority in the fact that "all *over the globe, negroes in the vicinity of another human race, are very soon subjected. On the contrary, not one of the other races has ever been enslaved by them*" (quoted in Virey, *Natural History*, 98).

23. Virey, *Natural History*, 19.

24. Nott and Gliddon, *Types of Mankind*, 458–59. *Types of Mankind* was only one of two monographs Nott and Gliddon published, with the other being the compendious and lavishly illustrated *Indigenous Races of the Earth; Or, New Chapters of Ethnological Inquiry* (Philadelphia: J. B. Lippincott, 1857).

25. While Morton (*Crania Americana*, 3) had suggested that an "all-wise Providence" had adapted each race "from the beginning to its peculiar local destination," the beginning referred to is after the great flood that left Noah and his family alone in the world. The claim for Nott's priority in the argument for separate creations is made in C. Loring Brace, "The 'Ethnology' of Josiah Clark Nott," *Bulletin of the New York Academy of Medicine* 50, no. 4 (1974): 516. The same claim is made in Horsman, *Josiah Nott*, 69.

26. Sir William Jones, "Discourse the Ninth, 'The Origin and Families of Nations'" (1792), in *The Works of Sir William Jones* (London: John Shore, 1807), 9:201, http://www.eliohs.unifi.it/testi/700/jones/Jones_Discourse_9.html.

27. Josiah C. Nott, *Two Lectures on the Natural History of the Caucasian and Negro Races* (Mobile, AL: Dade and Thompson, 1844), 3, 7, 41, https://www.google.com/books/edition/Two_Lectures_on_the_Natural_History_of_t/dJwE_A2viq8C?hl=en&gbpv=1&printsec=frontcover.

28. The comment on Nott's international reputation is from an 1866 article in an English journal quoted in Brace, "The 'Ethnology' of Josiah C. Nott," 515. As to his acquaintance with the leading scientific figures of the age, Nott wrote to a friend as early as 1848 saying, "I am now in Scientific correspondence with all Creation." Josiah C. Nott to Ephraim Squier, September 30, 1848, quoted in Horsman, *Josiah Nott*, 108. On February 14, 1849, Nott wrote to Squier, saying that his "great aim has been to get the world quarreling about niggerology and I have last succeeded and I think I shall sit on the fence now and enjoy the fight. . . . The subject attracts and all the little articles I have written on niggerology have been eagerly sought for in the South, and in the present excited state of the political world I think the thing will go well." Quoted in W. S. Jenkins, *Pro-Slavery Thought in the Old South* (Chapel Hill: University of North Carolina Press, 1985,) 260. On the state of scientific opinion in the United States in the 1840s and Nott's reputation as an ethnologist, see William Stanton, *The Leopard's Spots: Scientific Attitudes Toward Race in America 1815–59* (Chicago: University of Chicago Press, 1960); Horsman, *Josiah Nott*, 115–26; Reginald Horsman, *Race and Manifest Destiny: The Origins of American Racial Anglo-Saxonism* (Cambridge, MA: Harvard University Press, 1981).

29. Nott, *Two Lectures*, 28.

30. Ussher's much-ridiculed research is quite impressive in its way; one can see why it was considered a challenge to nineteenth-century ethnologists, who had to fill the gap between the first recorded traces of human existence and the origin of humanity as Ussher determined it. See https://www.propheticwitness.com/2019/02/the-Biblical-chronology-of-james-ussher.html.

31. As Terence D. Keel summarizes the situation, in which denial and evasion took the place of engagement, "During the early 19th-century professional geologists avoided altogether the question of human origins in order to stay above the partisan conflict between traditional chronologists and the secularizing concerns of eternalistic theories of the earth. . . . As a result, 19th-century geological claims about the deep antiquity of the earth had very little effect on how most 19th-century scholars viewed human history." Terence D. Keel, "Religion, Polygenism, and the Early Science of Human Origins," *History of the Human Sciences* 26, no. 2 (2013): 7, https://www.history.ucsb.edu/wp-content/uploads/History-of-the-Human-Sciences-2013-Keel-3-32.pdf.

32. See S. J. Wolfe, "Bringing Egypt to America: George Gliddon and the *Panorama of the Nile*," *Journal of Ancient Egyptian Interconnections* 8, no. 1 (2016): 1–20, http://jaei.library.arizona.edu/.

33. Stanton, *Leopard's Spots*, 46.

34. George R. Gliddon, *Ancient Egypt: Her Monuments, Hieroglyphics, History, Archæology, and Other Subjects Connected with Hieroglyphical Literature* (New York: J. Winchester, 1844), 44. See also George R. Gliddon, *Otia Ægyptiaca: Discourses on Egyptian Archeology* (Philadelphia: John Pennington, 1849). For a detailed discussion of Gliddon, Nott, and Morton on the question of Egypt as well as the management of Morton's reputation after his death, see Christopher R. Donohue, "Samuel Morton, Josiah Nott, and the Origins of the 'American School': Authority, Genius, and Systems-Building in Nineteenth-Century Ethnology" (MA thesis, University of Maryland, 2009), 74–93. For a wonderfully comprehensive account of Egypt as a commanding but ambivalent and unstable figure for the complexities associated with race in the United States, see Scott Trafton, *Egypt Land: Race and Nineteenth-Century Egyptomania* (Durham, NC: Duke University Press, 2004). See also John A. Wilson, *Signs and Wonders upon Pharoah: A History of American Egyptology* (Chicago: University of Chicago Press, 1964). On Gliddon, Morton, and Nott, see C. Loring Brace, *"Race" Is a Four-Letter Word: The Genesis of the Concept* (New York: Oxford University Press, 2005), 125–30. See also Ann Fabian, *The Skull Collectors: Race, Science, and America's Unburied Dead* (Chicago: University of Chicago Press, 2010), 103–12. Fabian's is an excellent account of the development, at once grisly and farcical, of American craniology; see especially chap. 3, 81–119.

35. At the time he wrote *Crania Americana*, Morton possessed 147 skulls. His collection eventually swelled to over eleven hundred, which have been preserved at the Penn Museum at the University of Pennsylvania. In 2024, the Museum began the process of "repatriating" and burying Morton's skulls. With additions, the total number in Penn's collection is over 1,300. See Graham Lee Brewer, "Penn Museum Buries the Bones of 19 Black Philadelphians, Causing a Dispute with Community Members," AP, February 15, 2014, https://apnews.com/article/morton-cranial-collection-penn-museum-bones-repatriation-8b87b5542d9dc18447f791ddfa87f121.

36. Morton rejected the notion that variations could be caused by environmental factors such as climate. To him, climate was but another word for chance, and it seemed both implausible and depressing that "chance, *chance* alone has caused all the physical disparity among men, from the noblest Caucasian form to the most degraded Australian and Hottentot." Quoted in Henry S. Patterson, "Memoir of the Life and Scientific Labors of Samuel George Morton," in Nott and Gliddon, *Types of Mankind*, xlix.

37. George Combe, "Appendix: Phrenological Remarks on the Relation Between the Natural Talents and Dispositions of Nations, and the Developments of Their Brain," in *Crania Americana, or, A Comparative View of the Skulls of Various Aboriginal Nations of North and South America*, by Samuel George Morton (Philadelphia: J. Dobson, 1839), 274. The following passage is representative of Combe's approach: "The size of the organs devoted to the moral sentiments may be estimated as follows. These organs lie in the coronal region of the head; and when the axis of the eye is parallel with the plane of the horizon, a horizontal line stretches across the forehead at the superior edges of the organs of causality and drawn backward till it touch the superior edges of the organs of cautiousness, would leave all

the moral organs above it, and the organs of the propensities common to man and the lower animals, below or behind it" ("Appendix," 279). On Combe, see Stanton, *Leopard's Spots*, 34–39. On phrenology, see Thomas F. Gossett, *Race: The History of an Idea in America* (1963; repr., New York: Oxford University Press, 1997), 70–77; Brace, *"Race" Is a Four-Letter Word*, 66–75. For a valuable account of the place of phrenology in the development of a global science, see James Poskett, *Materials of the Mind: Phrenology, Race, and the Global History of Science 1815–1920* (Chicago: University of Chicago Press, 2019).

38. Combe, "Appendix," 276.

39. See Marianne Sommer, "A Diagrammatics of Race: Samuel George Morton's 'American Golgotha' and the Contest for the Definition of the Young Field of Anthropology," *History of the Human Sciences* 37, no. 3–4 (2023), https://journals.sagepub.com/doi/10.1177/09526951221136771.

40. Morton, *Crania Americana*, 260.

41. Morton, *Crania Americana*, 4, 8, 27, 90–91, 55.

42. See James Cowles Prichard, *The Natural History of Man; Comprising Inquiries into the Modifying Influence of Physical and Moral Agencies on the Different Tribes of the Human Family* (London: H. Baillierre, 1843), 364. This appraisal of Morton has been widely denounced in recent years. In 1981, the historian of science Stephen Jay Gould reanalyzed Morton's data, reporting his results in *The Mismeasure of Man*, rev. ed. (New York: W. W. Norton & Company, 1976), 100–104. Gould found fault with some of Morton's numbers, but saw "no evidence of conscious fraud" inasmuch as Morton had scrupulously documented his methods and results, virtually inviting others to replicate his measurements. What Gould did allege was that Morton's "a priori convictions" concerning racial groups had resulted in a perhaps unwitting "finagling" of the evidence and a general "crudity" of interpretation. Interestingly, Gould himself was accused of finagling by John S. Michael, at the time an undergraduate student at Macalester College, who remeasured some of Morton's skulls in 1998 and found the original measurements accurate. And in 2018, Paul Wolff Mitchell considered unpublished data from Morton's experiments that further confirmed Morton's results. Regardless of any methodological deficiencies or unconscious biases in Morton's work, however, there were other problems with Morton's project. The number of skulls was far from adequate to draw conclusions about large (and poorly defined) groups, nor were the skulls identified by sex, age, or crucially, body size. The most fundamental issue with Morton's project, however, was its basic premise, that intelligence, moral character, or anything can be correlated in any way with cranial capacity. For an indictment of Morton for conscious as well as unconscious bias in his methods of generating statistics, see Paul Wolff Mitchell and John S. Michael, "Bias, Brains, and Skulls: Tracing the Legacy of Scientific Racism in the Nineteenth-Century Works of Samuel George Morton and Friedrich Tiedemann," in *Embodied Difference: Divergent Bodies in Public Discourse*, ed. Jamie A. Thomas and Christina Jackson (Lanham, MD: Rowman & Littlefield, 2019), 77–98. For a sophisticated and thorough defense of Morton against the accusation of unconscious racism by Gould, see Jason Lewis, David DeGusta, Marc R. Meyer, Janet Monge, Alan E. Mann, and Ralph L. Holloway, "The Mismeasure of Science: Stephen Jay Gould Versus Samuel George Morton on Skulls and

Bias," *PLOS Biology* (July 2011), https://doi.org/10.1371/annotation/138c7c99-249f-432c-a4f7-2993b7b87c0a.

43. Morton, *Crania Americana*, 30. The authority is Dominique-Vivant Denon, *Voyages dans la Basse et la Haute Egypt pendant las campagnes du General Bonaparte* (1802), 2:296. As Morton noted in *Crania Ægyptiaca*, red had a symbolic significance and was used in paintings across the whole nation. *Crania Ægyptiaca or Observations on Egyptian Ethnography Derived from Anatomy, History and the Monuments* (Philadelphia: John Penington, 1844), 28.

44. Morton, *Crania Americana*, 31, 27, 29.

45. For a detailed tracing of all the ways arguments about race in the United States were routed through the example of Egypt, see Trafton, *Egypt Land*, esp. 41–84.

46. Morton, *Crania Americana*, 251, quoted in Petrus Camper, *Dissertation sur les variétés naturelles*, 42. The reference to Blumenbach is in Morton, *Crania Americana*, 140.

47. Morton, *Crania Americana*, 7.

48. Nott, *Two Lectures*, 24–25.

49. Samuel George Morton, *Brief Remarks on the Diversities of the Human Species* (Philadelphia: Merrihew & Thompson, 1842), 6, 21, 22.

50. Samuel George Morton, *Crania Ægyptiaca*, 3. Most of the book's text is taken up with ethnological descriptions of the nine groups living in Egypt.

51. Morton, *Crania Ægyptiaca*, 37, 61–62, 66.

52. Combe's premise about the correlation of brain size and intelligence would reappear in Nott and Gliddon, *Types of Mankind* (463): "brains below a certain size are always indicative of idiocy, and that men of distinguished mental faculties have large heads."

53. Morton, *Brief Remarks*, 6.

54. Morton, *Crania Americana*, 81, 82. In each edition of *Crania Americana* that I have seen, the figure for the American in the famous footnote appears to be eighty-two cubic inches, altered to eighty by a mark that seems to have been made by non-mechanical means. The difference meant that Americans were ranked below, rather than above, Malays.

55. Morton, *Crania Americana*, 88.

56. Morton, *Crania Americana*, 2, 152.

57. Morton, *Brief Remarks*, 2–3. In the footnote to these words, Morton (*Crania Americana*, 2) cites a long passage from Hugh Murray, *Encyclopedia of Geography*, in which Murray marvels at the "remarkable analogy" between the predictions made by Noah about his three sons and "the actual state of those races which are generally supposed to have sprung from them."

58. Morton, *Brief Remarks*, 8.

59. Morton, *Brief Remarks*, 7; Morton, *Crania Ægyptiaca*, 38.

60. The correspondence began in 1845, when Morton, having read Nott's lecture "The Mulatto a Hybrid," wrote to Nott in appreciative terms, including some articles of his own (and requesting any alligator skulls Nott might possess). On the Nott-Morton relationship, see Patterson, "Memoir," xlix–1; Horsman, *Josiah Nott*, 94–107. See also Brace, *"Race" Is a Four-Letter Word*, 114–17; Paul A. Erickson, "The Anthropology of

Josiah Clark Nott," *Kroeber Anthropological Society Papers* (1986): 103–20, https://digitalassets.lib.berkeley.edu/anthpubs/ucb/text/kas065_066-013.pdf.

61. For "original diversity," see Samuel George Morton, letter to George R. Gliddon, April 27, 1851, quoted in Patterson, "Memoir," li. For "primeval diversities," see Samuel George Morton, *Hybridity in Animals and Plants, Considered in Reference to the Question of the Unity of the Human Species* (New Haven, CT: B. Hamlin, printer to Yale College, 1847), 4. For "primordial organic forms," see Samuel G. Morton, *Proceedings of the Academy of Natural Sciences* (Philadelphia, 1850), 82, quoted in Nott and Gliddon, *Types of Mankind*, 375.

62. Quoted in Patterson, "Memoir," l.

63. Bachman, a proslavery (and slaveholding) Lutheran minister and professor of natural history at the College of Charleston, was Morton and Nott's most persistent, vigorous, and unreconciled antagonist. He agreed wholeheartedly with the premise of Negro inferiority, but defended Biblical monogenesis as consistent with science, and even argued that given the common origins of humankind, the entire concept of race should be abandoned. For a discussion of Bachman's running disputes with Morton and Nott, see Stanton, *Leopard's Spots*, esp. 123–36, 170–73; Brace, *"Race" Is a Four-Letter Word*, 106–24; Brad D. Hume, "Quantifying Characters: Polygenist Anthropologists and the Hardening of Heredity," *Journal of the History of Biology* 41, no. 1 (2008): 119–58. In the course of his extended campaign against Morton and Nott, Bachman wrote many letters and essays and one book, *The Doctrine of the Unity of the Human Races, Examined on the Principles of Science* (1850). His lengthy review of *Types of Mankind* was published as a book: John Bachman, *Notice of the "Types of Mankind"* (Charleston, SC: James, Williams and Gitsinger, 1854). Bachman was a friend of John James Audubon and collaborated with him on *The Viviparous Quadrupeds of North America* (Philadelphia: J. T. Bowen, 1845–48).

64. Samuel George Morton, letter to Josiah C. Nott, January 29, 1850, quoted in Patterson, "Memoir," l.

65. Samuel George Morton, "Notes on Hybridity," letter to the editor, *Charleston Medical Journal and Review*, December 15, 1850, 145, https://digirepo.nlm.nih.gov/ext/mhl/101206310/PDF/101206310.pdf.

66. Samuel George Morton, "Notes on Hybridity, Designed as a Further Supplement to a Memoir on That Subject in a Former Number of This Journal," *Charleston Medical Journal and Review* (January 31, 1851): 8, https://wellcomecollection.org/works/tat796sa. The quote is from W. F. Van Amringe, *An Investigation of the Theories of the Natural History of Man, by Lawrence, Prichard, and Others* (New York: Baker & Scribner, 1848), 421. Morton's reasons for embracing polygenesis were complex. Several months before his death, Morton wrote to Nott, saying something he never said in any publication: that one reason he had finally embraced polygenesis was that he realized that the alternative—fidelity to the Biblical account of a single human family issuing from Adam and Eve—implied that the human race had begun in incest, a prospect that held, for Morton, "the clue to all ungodliness." Morton even confessed to having composed a couplet that expressed his concerns:

"*Q*. Whence came that curse we call primeval sin?"
"*A*. From Adam's children breeding in and in."

Quoted by Nott in Nott and Gliddon, *Types of Mankind*, 409. Nott had used the phrase "in and in" (407).

67. Josiah C. Nott, letter to Samuel George Morton, June 27, 1847, Samuel George Morton Papers, Library Company of Philadelphia. Nott's biographer concurs, saying that by the late 1840s, Nott's work was received "as the major contribution of a man of science," its reputation enhanced by Nott's "emphasis on broad racial themes rather than on slavery itself" and his insistence on the independence of science from clerical interpretations of the Bible. Horsman, *Josiah Nott*, 102.

68. R. W. Gibbs, *Charleston Medical Journal* (1851), quoted in Stanton, *Leopard's Spots*, 144.

69. Patterson, "Memoir," xxv.

70. Nott and Gliddon, *Types of Mankind*, 246, 255, 260.

71. "It will be seen," Nott wrote, "that we recognize no substantial difference between the terms *types* and species" (Nott and Gliddon, *Types of Mankind*, 81). One definition of a "type" would be a hypothetical individual who possesses all characteristics equal to the averages of all the individuals in the sample. To this Nott contributed other elements, which Paul A. Erickson describes: "Some of the key features of his medical anthropology include physiological vitalism, his racial typology and idea of permanence of racial types, his disbelief that races acclimatize or are transformed by nature, and his lack of scientific experimentation. All of these imply a kind of Platonic essentialism, in which life is separate from nature and some kinds of organisms are separate from others. Such separation," Erickson says, "is foreign to the modern ethos of science, in which all organic and inorganic things are interconnected." Erickson, "The Anthropology of Josiah Clark Nott," 115. In the introduction to *Types of Mankind*, Nott de-emphasized plural creation in favor of the permanent and "radical differences of humanity's races" as they presently exist, citing in support "the mastermind of JOHN C. CALHOUN" (Nott and Gliddon, *Types of Mankind*, 50).

72. For contemporary responses to *Types of Mankind*, including the outraged Presbyterian reaction, see Stanton, *Leopard's Spots*, 163–73. For other accounts of the American School, see Gossett, *Race*, 54–83; Brace, "The 'Ethnology' of Josiah Clark Nott"; Brace, *"Race" Is a Four-Letter Word*, 76–177.

73. Nott and Gliddon, *Types of Mankind*, 74, 79.

74. Morton, *Crania Americana*, 5; Nott and Gliddon, *Types of Mankind*, 79.

75. Nott and Gliddon, *Types of Mankind*, 407.

76. Morton, *Crania Americana*, 7.

77. Nott and Gliddon, *Types of Mankind*, 409.

78. Nott and Gliddon, *Types of Mankind*, 404.

79. George M. Fredrickson, *The Black Image in the White Mind: The Debate on Afro-American Character and Destiny, 1817–914* (1971; repr., Middletown, CT: Wesleyan University Press, 1987), 79; Horsman, *Josiah Nott*, 92; Brace, "The 'Ethnology' of Josiah Clarke Nott," 516. For an overview of the "racial politics" of the American School, see Lee D. Baker, *Anthropology and the Racial Politics of Culture* (Durham, NC: Duke University Press, 2010).

80. Keel, "Religion, Polygenism, and the Early Science of Human Origins," 10. Keel also stresses Nott's Christianity, arguing that he relied, often without saying as much, on ideas about the order of nature derived from Christian natural theology.

For a full exploration of the derivation of nineteenth-century racial science from premodern Christian thought, see Terence D. Keel, *Divine Variations: How Christian Thought Became Racial Science* (Palo Alto, CA: Stanford University Press, 2020).

81. William Howard Russell, *My Diary North and South*, ed. Fletcher Pratt (New York, 1954), 126–27, quoted in Horsman, *Josiah Nott*, 258.

82. Stanton, Leopard's Spots, 160.

83. J. C. Nott, "Geographical Distribution of Animals and the Races of Man," *New Orleans Medical and Surgical Journal* 9 (1853): 732, 733.

84. Nott, *Two Lectures*, 41.

85. "Death of Samuel George Morton, M. D.," *Medical Examiner and Record of Medical Science*, June 1851, 385, quoted Fabian, *Skull Collectors*, 223.

86. Patterson, "Memoir," xxxii.

87. Stanton, *Leopard's Spots*, 35.

88. Stanton, *Leopard's Spots*, 38. The colleague was the Swedish anatomist Anders Retzius, who developed the "cephalic index," achieved by dividing the length of the skull by the breadth. Retzius coined the terms dolicocephalic for long, narrow skulls with an index of eighty or less, and brachycephalic for wider skulls.

89. George Combe, *Notes on the United States of North America During a Phrenological Visit in 1838-9-40* (Edinburgh: Maclachlan, Stewart & Co., 1841), 1:254. In this text, Combe also recorded his disgust for the racial prejudice exhibited in the North.

90. Combe, *Notes*, 2:112.

91. Combe, "Appendix," 282.

92. Nott, *Two Lectures*, 23.

93. This mighty aggregation, dedicated to "the venerable and universally celebrated Blumenbach," originated in a single volume titled *De generis humani varietate* (On the origins of human varieties), published in 1808 by the author at the age of twenty-two. This was transformed into the two-volume *Researches into the Physical History of Man* (1813), and then into three lavishly illustrated and increasingly lengthy editions with plates, maps, and atlases that bore the title of *Researches into the Physical History of Mankind* (1826, 1836–47, 1841–47). One of the most significant works of physical anthropology published in the first half of the nineteenth century, Prichard's text is commonly regarded as the most influential and detailed pre-Darwinian account of human variation. It is in this work that race appears decisively, and perhaps for the first time, not as a set of ethnological facts to be described but as a scientific problem inviting research and awaiting solution.

94. James Cowles Prichard, *Researches into the Physical History of Mankind*, 4th ed. (London: Sherwood, Gilbert, and Piper, 1841), 1:220, 221.

95. James Cowles Prichard, *Researches into the Physical History of Mankind*, 2nd ed. (London: John and Arthur Arch, 1826), 2:268.

96. Prichard, *Researches* (1941), 1:177–82.

97. Nott and Gliddon, *Types of Mankind*, 54–55.

98. Nott and Gliddon, *Types of Mankind*, 54–57; see also 189–91.

99. Patterson, "Memoir," xvii.

100. Patterson, "Memoir," lii.

101. Brace describes Agassiz as "an orthodox monogenist" and conventional Christian prior to visiting Morton in the United States in 1846. Brace, *"Race" Is a Four-Letter Word*, 97.

102. Translated and quoted in Edward Lurie, *Louis Agassiz: A Life in Science* (Chicago: University of Chicago Press, 1960), 256.

103. Agassiz's remarks were published as "A Correction of Some Mis-statements," *Proceedings, AAAS* III (1850) 3:106–7, quoted in Stanton, *Leopard's Spots*, 149. For an account of the incident, see Horsman, *Josiah Nott*, 114–15.

104. Brace, *"Race" Is a Four-Letter Word*, 101.

105. Louis Agassiz, "Sketch of the Natural Provinces of the Animal World and Their Relation to the Different Types of Man," in Nott and Gliddon, *Types of Mankind*, lxxiv–lxxv.

106. Agassiz, "Sketch," lxxvi.

107. Brace, *"Race" Is a Four-Letter Word*, 103.

108. Stanton, *Leopard's Spots*, 35.

109. Le Comte de Gobineau, *Essai sur l'inégalité des races humaines* [Essay on the inequality of human races], 4 vols. (Paris: Librairie Firmin Didot, 1853–55).

110. Hannah Arendt, *The Origins of Totalitarianism* (Orlando, FL: Harvest Book, 1976), 172. In an 1856 letter, Gobineau told a friend that his book had grown not out of race-hatred but out of his "hatred for democracy and its weapon, the Revolution." Quoted in Michael D. Biddiss, *Father of Racist Ideology: The Social and Political Thought of Count Gobineau* (New York: Weybright and Talley, 1970), 514.

111. J. C. Nott, "Appendix," in *The Moral and Intellectual Diversity of Races*, by Count A. de Gobineau, trans. Henry Hotz (Philadelphia: J. B. Lippincott & Co., 1856), 463.

112. Henry Hotz, "Analytical Introduction," in *Moral and Intellectual Diversity of Races*, by Count A. de Gobineau, trans. Henry Hotz (Philadelphia: J. B. Lippincott & Co., 1856), 32.

113. Hotz, "Analytical Introduction," 65, 99.

114. Henry Hotz, "Introductory Note," in *Moral and Intellectual Diversity of Races*, by Count A. de Gobineau, trans. Henry Hotz (Philadelphia: J. B. Lippincott & Co., 1856), 243.

115. Gobineau, *Essai*, 852–53, in Biddiss, *Father of Racist Ideology*, 144.

116. Hotz, "Introductory Note," 241.

117. "Ce nom d'Anglo-Saxons parait flatter l'imagination des habitants de la grande confédération transatlantique; malgré le droit de plus en plus équivoque que la population actuelle peut avoir à le réclamer." Gobineau, *Essai*, Book 6, Chap. 7, 315 (my translation; only Book 1 has been translated into English), https://classiques.uqam.ca/classiques/gobineau/essai_inegalite_races/essai_inegalite_races_2.pdf.

118. Gobineau, *Essai*, book 2, chap. 8, 301 (my translation). The relevant passage reads, "Certainement l'élément noir est indispensable pour développer le génie artistique dans une race, parce que nous avons vu quelle profusion de feu, de flammes, d'étincelles, d'entraînement, d'irréflexion réside dans son essence, et combien l'imagination, ce reflet de la sensualité, et toutes les appétitions vers la matière le rendent propre à subir les impressions que produisent les arts, dans un degré d'intensité tout à fait inconnu aux autres familles humaine."

119. Gobineau, *Essai*, book 1, chap. 16, 198 (my translation). The original reads, "de merveilles auxquelles il faut applauder."

120. Hotz, "Introductory Note," 241.

121. See Gobineau, *Essai*, book 6, chap. 8. See also Lonnie A. Burnett, "Introduction: Henry Hotze, Confederate Propagandist," in *Henry Hötze, Confederate Propagandist: Selected Writings on Revolution, Recognition, and Race*, ed. Lonnie A. Burnett (Tuscaloosa: University of Alabama Press, 2008), 5. See also the informative essay by Michelle M. Wright, "Nigger Peasants from France: Missing Translations of American Anxieties on Race and the Nation," *Callaloo* 24 (Autumn 1999): 831–52, https://www.jstor.org/stable/3299815.

122. Frederick Douglass, "The Claims of the Negro, Ethnologically Considered," in *The Life and Writings of Frederick Douglass*, ed. Philip S. Foner (New York: International Publishers, 1950), 2:289–309.

123. Frederick Douglass, "The End of All Compromises with Slavery—Now and Forever," in *Life and Writings of Frederick Douglass*, ed. Philip S. Foner (New York: International Publishers, 1950), 2:282, 284.

124. Philip S. Foner, *Frederick Douglass: A Biography* (New York: Citadel Press, 1969); David W. Blight, *Frederick Douglass: Prophet of Freedom* (New York: Simon & Schuster, 2018). For a methodical consideration of the philosophical claims made by Douglass in this address, see Roderick M. Stewart, "The Claims of Frederick Douglass Philosophically Considered," in *Frederick Douglass: A Critical Reader*, ed. Bill E. Lawson and Frank M. Kirkland (Malden, MA: Blackwell, 1990), 145–72.

125. Douglass, "Claims of the Negro," 295. "Smith" is Charles Hamilton Smith, author of *The Natural History of the Human Species: Its Typical Forms, Primeval Distribution, Filiations, and Migrations* (Boston: Gould & Lincoln, 1851). Citing Buffon and Cuvier, Smith had endorsed Morton's craniology, attributed Negro slavery to "a curse, as old as the origin of society," and described Caucasians as beautiful, decisive, graceful, physically powerful, morally profound, and intellectually brilliant. He also affirmed Nott's argument about the infertility of mulattoes, saying that a crossing between "two such dissimilar forms of Man, as a typical African negro and an European," with no intervening admixture, "would, in all probability, end in . . . sterility, or at least in sensible degradation" (Smith, *Natural History*, 227, 171).

126. Douglass, "Claims of the Negro," 289, 292, 291.

127. Douglass frequently compares the dehumanized status of the slave to that of a "brute" or "beast." In a passage included as an appendix to his *My Bondage and My Freedom*, to take just one among many examples, he writes, "I hear the doleful wail of fettered humanity, on the way to the slave markets, where the victims are to be sold like horses, sheep, and swine, knocked off to the highest bidder." The passage, given the title of "The Internal Slave Trade," was taken from Douglass's July 5, 1851, speech at Rochester, New York; it has been published in full under other names, including "What to the Slave Is the Fourth of July?" See Frederick Douglass, *Autobiographies* (New York: Library of America, 1994), 438.

128. Douglass, "Claims of the Negro," 290.

129. Douglass, "Claims of the Negro," 292.

130. Douglass, "Claims of the Negro," 293, 307.

131. Douglass, "Claims of the Negro," 292.

132. Douglass, "Claims of the Negro," 307.

133. Frederick Douglass, "God Almighty Made but One Race," interview, January 25, 1884, in *The Frederick Douglass Papers*, ed. John W. Blassingame (New Haven, CT: Yale University Press, 1992), series 1, 5:147.

134. Frederick Douglass, "Abraham Lincoln—The Great Man of Our Century," speech delivered in Brooklyn, NY, February 13, 1893, in *The Frederick Douglass Papers*, ed. John W. Blassingame (New Haven, CT: Yale University Press, 1992), series 1, 5:537.

135. James McCune Smith, introduction to *My Bondage and My Freedom*, by Frederick Douglass (1855; repr., London: Penguin Classics, 2003), 132, 137.

136. Frederick Douglass, *My Bondage and My Freedom* (1855; repr., London: Penguin Classics, 2003), 304.

137. All three of Douglass's autobiographies, including the third, *Life and Times of Frederick Douglass* (1893), are included in Douglass, *Autobiographies*.

138. Douglass, *Autobiographies*, 53, 286.

139. Josiah C. Nott, *The Instincts of Races* (New Orleans: L. Graham, 1866), 4. This fascinating postwar pamphlet represents one of Nott's final ethnological deliverances. It begins by describing "*the instinctive dislike of the negro race to agricultural labor*" as a settled fact that justified the compulsory aspect of slavery, predicts the extinction of black people in the United States, and rehearses his arguments about races and species, supported as usual with evidence from ancient Egypt (1). As for Darwin, his less controversial *The Descent of Man, and Selection in Relation to Sex* (London: John Murray, 1871), while speaking of "the so-called races of man," rejected the notion of plural creations and argued for common descent from "a single primitive stock" (1:3, 229). In this book Darwin referred to Caucasians as "more civilized," and "the negro or Australian" as more proximate to primates (1:201).

140. Quoted from *The New York Age* in Blight, *Frederick Douglass*, 702.

141. Foner gives a detailed account of the negotiations in *Frederick Douglass*, 350–61. In the course of their negotiations, Firmin may have told Douglass of his own doubts about the scientific validity of the race concept; he almost certainly told Douglass of his long-standing admiration for him. In *De l'égalité des races humaines* (1885), translated by Asselin Charles as *The Equality of the Human Races* (Chicago: University of Chicago Press, 2002), Firmin had described Douglass as "one of the most striking illustrations of the concept of equality of the human races" (271). In the same book, Firmin also wrote in a spirit of deep admiration about *Narrative of the Life of Frederick Douglass* (331–33).

142. Frederick Douglass, "The Lessons of the Hour" (also sometimes called "Why the Negro Is Lynched"), in *The Frederick Douglass Papers*, ed. John W. Blassingame (New Haven, CT: Yale University Press, 1992), series 1, 5:576–77. See also Blight, *Frederick Douglass*, 740–48.

143. Horsman, *Josiah Nott*, 321.

144. For an account of Nott's final deliverances on the subject of race, see Brace, "The 'Ethnology' of Josiah Clark Nott," 527–28, 516.

145. Horsman, *Josiah Nott*, 336.

146. Paul Broca, "Mémoires sur l'Hybridité," in *Mémoires d'Anthropologie* (Paris: C. Reinwald, 1877), 3:321–623; see also 339, 493, 525–31.

147. *Mobile Register*, April 1, 1873, quoted in Horsman, *Josiah Nott*, 335. In 2020, Nott's name was removed from a building on the University of Alabama campus; the building was renamed Honors Hall.

148. On the Lost Cause, see Edward A. Pollard, *The Lost Cause: A New Southern History of the War of the Confederates* (1866; repr., War College Series, 2015); Edward A. Pollard, *The Lost Cause Regained* (1868; repr., Scholarly Publishing Office, University of Michigan Library, 2005). See also Charles Reagan Wilson, *Baptized in Blood: The Religion of the Lost Cause, 1865–1920* (Athens: University of Georgia Press, 2009). On the backlash to Reconstruction, see Henry Louis Gates Jr., *Stony the Road: Reconstruction, White Supremacy, and the Rise of Jim Crow* (New York: Penguin Press, 2019), esp. 1–54. On *The Birth of a Nation*, see Geoffrey Galt Harpham, *Citizenship on Catfish Row: Race and Nation in American Popular Culture* (Columbia: University of South Carolina Press, 2023), 15–68.

Chapter Four: Equal or Unequal?

1. In volume 2 of *Mémoires de la Société d'anthropologie de Paris*, Broca had prepared an instructional guide for aspiring craniologists, *Instructions craniologioques et craniométriques* (Paris: Librairie Georges Masson, 1875). On Broca and the founding of the Anthropological Society, see C. Loring Brace, *"Race" Is a Four-Letter Word: The Genesis of the Concept* (New York: Oxford University Press, 2005), 144–58.

2. Jules Michelet, *History of France*, trans. G. H. Smith (New York: D. Appleton and Company, 1882), 1:67.

3. For a colorful account of the history of the Society, see Jean-Claude Wartelle, "La Société d'Anthropologie de Paris de 1859 à 1920" [The Anthropology Society of Paris from 1859–1920], 13–24, https://www.cairn.info/revue-histoire-des-sciences-humaines-2004-1-page-125.htm, originally published in *Revue d'Histoire des Sciences Humaines* 1, no. 10 (2004): 125–71. The occasion referred to occurred during a discussion in July 1874 of the drop in birth rates in France. Royer, the only woman admitted to the Society, had pointed to a long list of social and political conditions that affected birth rates, introducing as relevant considerations several comments on female sexuality, and describing in detail the sexual pleasures offered by mistresses to their lovers. Royer was not present at the 1884 meeting evoked in what follows here, but she was present at the April 7 and April 21, 1892, meetings that Firmin attended.

4. On this arrangement, see Jean Price-Mars, *Joseph Anténor Firmin* (Port-au-Prince: Imp. Séminaire adventiste, 1964), 136–46.

5. From Jules François Camille Ferry, "Speech Before the French Chamber of Deputies, March 28, 1884," in *Discours et Opinions de Jules Ferry*, ed. Paul Robiquet (Paris: Armand Colin & Co., 1897), 215–18. Translated by Ruth Kleinman in Brooklyn College Core Four Sourcebook. For the original post, see "Internet Modern History Sourcebook," Fordham University, https://sourcebooks.fordham.edu/mod/1884ferry.asp. For an early account of the ideology of French imperialism, see Herbert Ingram Priestley, *France Overseas: A Study of Modern Imperialism* (London: Routledge, 1938). See also Matthew Burrows, "'*Mission civilisatrice*': French Cultural Policy in the Middle East, 1860–1914," *Historical Journal* 29, no. 1 (1986): 109–35.

6. The agenda of this meeting is recorded in *Bulletins de la Société d'Anthropologie de Paris*, 5me Fascicule (1884), *Nature* 32 (1885): 165–66, https://www.nature.com/articles/032165b0.

7. Paul Broca, *Mémoires d'anthropologie*, 4 vols. (Paris: C. Reinwald, 1871–78). A fifth volume was published in 1888, eight years after Broca's death.

8. Broca, *Mémoires*, vol. 1. On the "auricular point," see Broca, *Mémoires*, 1:80.

9. "Le nom de races, au contraire, peut être adopté par tout le monde, et c'est pour cela qu'il a maintenant prévalu." Paul Broca, "D'Anthropologie," in *Mémoires*, 1:9. The translations into English are mine unless otherwise indicated. Prichard gave the same reason for using the word race as a term that could accommodate both species and variety. See James Cowles Prichard, *Researches into the Physical History of Mankind*, 4th ed. (London: John and Arthur Arch, 1826), 1:91.

10. Paul Broca, "Mémoires sur l'Hybridité," in *Mémoires*, 3:321–623.

11. Paul Broca, "Recherches sur l'Hybridité Animale en Général et sur l'Hybridité Humaine en Particulier," in *Mémoires*, 3:321–568. This immense work constitutes most of "Mémoires sur l'Hybridité."

12. Broca, "Mémoires sur l'Hybridité," 3:330.

13. Broca had proposed a research project founded on this premise to the Société de Biologie—which, however, rejected it in the strongest possible terms as an affront to the Biblical story of creation. Taking this rejection as an affront to himself, Broca (*Mémoires*, 3:323–25) had immediately left that group to found the Anthropology Society.

14. Broca, *Mémoires*, 3:417.

15. "Il y eut un moment où le débat politique parut se concentrer sur ce terrain scientifique, où l'on put croire que la destinée des nègres dépendait de l'opinion des législateurs sur l'influence nigrifîante du soleil d'Afrique." Broca, *Mémoires*, 2:497.

16. Paul Broca, *On the Phenomena of Hybridity in the Genus Homo* (London: Roberts, and Green, 1864), 21, 30. This translation was produced for the Anthropological Society of London led by Hunt. In Broca's *Mémoires*, this text is "Des Phénomènes d'Hybridité dans le Genre Humaine," 3:493–567. All translations of this text are taken from the London edition, *Phenomena*, https://www.gutenberg.org/files/47137/47137-h/47137-h.htm.

17. One polygenist, Knox, had said, "I do not believe that any mulatto race can be maintained beyond the third or fourth generation by *mulattoes merely*; they must intermarry with the pure races, or perish." Robert Knox, *The Races of Men* (London: Henry Renshaw, 1850), 156. For a French statement along the same lines, see Georges Pouchet, *De la pluralité des races humaines; essai anthropologique* (Paris: J. B. Baillière et fils, 1858), esp. 134–53. Pouchet cites among his many authorities Morton, Nott, and Gliddon. For a vigorous assertion of this argument, see Nott and Gliddon, *Types of Mankind*, chap. 12.

18. Broca, *Phenomena*, 30.

19. Étienne Serres, *Rapport sur les résultats scientifiques du voyage de circumnavigation de l'Astrolabe et de la Zélée*, in *Comptes rendus*, 13:648, quoted in Broca, *Phenomena*, 28. For the the French original, see Broca, "Des Phénomènes d'Hybridité dans le Genre Humaine," 3:521–22. In his massive tribute to Broca, *L'Anthropologie*,

Topinard proposes as a refinement the judgment that the Negro's penis is "longer and more voluminous in a flaccid state" than the white man's, but "the reverse is true when it is in a state of erection" ("le pénis du nègre dans l'état de flaccidité est plus long et plus volumineux que celui du blanc; à l'état d'érection il semble que ce soit l'inverse"). Paul Topinard, *L'Anthropologie* (Paris: C. Reinwald, 1876), 381, translated by Robert T. H. Bartley as *Anthropology* (London: Chapman and Hall, 1878).

20. Broca, *Phenomena*, 29.

21. Nott and Gliddon, *Types of Mankind*, 407, quoted in Broca, *Mémoires*, 1:279: "Through the operation of the laws of hybridity alone, the human family might possibly become exterminated by a thorough amalgamation of all the various types of mankind now existing on earth."

22. Paul Broca, "Phenomenon of Hybridity," in *Mémoires*, 3:526. Broca attributes this list to an 1842 publication by Nott whose title he cannot recall, although he may be referring to an unattributed essay by Nott in the *Philadelphia Medical Journal*. All these points and others are, however, reproduced in Josiah C. Nott, *Two Lectures on the Natural History of the Caucasian and Negro Races* (Mobile, AL: Dade and Thompson, 1844), 30–35.

23. "Le pied du nègre rappelle un peu la main postérieure des quadrumane." Broca, *Mémoires*, 3:397.

24. Broca, *Phenomena*, 30.

25. Broca, *Phenomena*, 44–45, 71.

26. "Je laisse donc mon livre tel que je l'ai fait et je n'y changerai absolument rien" (I leave my book as it is and I will change absolutely nothing). Arthur comte de Gobineau, "Avant-propos de la deuxième edition," in *Essai sur l'inégalité des races humaines* [Essay on the inequality of human races], ed. Hubert Juin (1853–55; repr., Paris: Belfond, 1967), 36; https://francegenocidetutsi.org/GobineauEssaiInegaliteRaces1.pdf. This online edition contains four of Gobineau's six books; for books 5 and 6, see http://tybbot.free.fr/Tybbow/Livres/Gobineau/essai_inegalite_races_2.pdf. Translations are from Arthur, comte de Gobineau, *The Inequality of Human Races*, trans. Adrian Collins (London: Heinemann, 1915). The bibliography on Gobineau is extensive, but for a comprehensive introduction to his thinking, see Michael D. Biddiss, *Father of Racist Ideology: The Social and Political Thought of Count Gobineau* (New York: Weybright and Talley, 1970).

27. While the first two sections of this chapter are narrative and therefore fictional in form, they are based on indications and inferences from Anténor Firmin, *The Equality of the Human Races*, trans. Asselin Charles, intro. Carolyn Fluehr-Lobban (Urbana: University of Illinois Press, 2000), originally published as *De l'égalité des races humaines* (Paris: Librairie Cotillon, 1885). Most of the quotations or arguments of Broca referred to in those parts are attributed either directly or indirectly to Broca by Firmin in this text. Other comments are taken directly from Firmin's text. For extended comparisons between "girls of the tropics" and Parisian women, see Firmin, *Equality*, 181–94. On the effect of "frequency of usage" on penis size, see Firmin, *Equality*, 67. On Firmin's misogyny, see the "Second to No Other Race" section in this chapter. On the Anthropology Society ignoring Gobineau personally but following him intellectually, see Firmin, *Equality*, 145. Focusing on Firmin, I have ignored

the fact that Louis-Joseph Janvier, a fellow Haitian even younger than Firmin, had been admitted to the Anthropology Society before Firmin and had proposed Firmin's membership. In 1884, Janvier published a hundred-page pamphlet, more political than scientific, with the title *L'Égalité des Races* (Paris, 1884), dedicated to "A la jeune Haiti." See https://www.google.com/books/edition/L_%C3%A9galit%C3%A9_des_races/nKxDAQAAMAAJ?hl=en&gbpv=1&printsec=frontcover. This pamphlet had appeared the previous year in the *Revue de la Jeune France* (August 1883) with the title "M. Renan et l'Égalité des races: Bretons et Nègres." Janvier does not mention either Gobineau or Broca.

28. Gobineau, *Inequality*, xiv. As Arendt commented, "Nobody before Gobineau thought of finding one single reason, one single force according to which civilization always and everywhere rises and falls." Hannah Arendt, *The Origins of Totalitarianism* (1968; repr., Orlando, FL: Harvest Book, 1976), 171.

29. Gobineau, *Inequality*, 210.

30. For Gobineau's equivocations on the subject of original unity or multiplicity, see *Inequality*, esp. 117–40.

31. Gobineau, *Inequality*, 119.

32. Gobineau, *Inequality*, 205–6, 180–81.

33. Gobineau, *Inequality*, 206.

34. Gobineau, *Inequality*, 206.

35. Gobineau, *Inequality*, 207.

36. Gobineau, *Inequality*, 207, 151.

37. Gobineau, *Inequality*, 58.

38. Gobineau, *Inequality*, 45.

39. Gobineau, *Inequality*, 48.

40. Gobineau, *Inequality*, 48–49.

41. Gobineau, *Inequality*, 49.

42. The phrase is from Price-Mars, *Joseph Anténor Firmin*, 114. As Price-Mars describes the Negro-mulatto struggles in the 1883 civil war that had resulted in Firmin's posting to Paris, "La couleur de la peau des principaux acteurs qui menaient la danse est venue apporter un assaisonnement terriblement pimenté à cette bouillabaisse de prétentions, de mensonges et d'ambitions" (The color of the skin of the main actors who led the dance brought a terribly spicy seasoning to this bouillabaisse of pretensions, lies, and ambitions) (35).

43. For a comprehensive discussion, see Léon Poliakov, *The Aryan Myth: A History of Racist and Nationalist Ideas in Europe*, trans. Edmund Howard (New York: Basic Books, 1971), esp. 225–54. See also Colin Kidd, *The Forging of Races: Race and Scripture in the Protestant Atlantic World, 1600–2000* (Cambridge: Cambridge University Press, 2006), esp. 168–202. Kidd focuses primarily on the work of nineteenth-century historian of religion Max Müller and others with a similar orientation.

44. Gobineau, *Inequality*, xv, 210.

45. Gobineau, *Essai*, 6:335, http://tybbot.free.fr/Tybbow/Livres/Gobineau/essai_inegalite_races_2.pdf. "L'espèce blanche, considérée abstractivement, a désormais disparu de la face du monde. Après avoir passé l'âge des dieux, où elle était absolument pure; l'âge des héros, où les mélanges étaient modérés de force et de nombre;

l'âge des noblesses, où des facultés, grandes encore, n'étaient plus renouvelées par des sources taries, elle s'est acheminée plus ou moins promptement, suivant les lieux, vers la confusion définitive de tous ses principes, par suite de ses hymens hétérogènes." Quoted in Firmin, *Equality*, 438–39.

46. As Firmin (*Equality*, 116) points out, "The term Caucasian is authoritative simply because it is old, for the fact is that several distinct races inhabit the Caucasus."

47. Gobineau, *Inequality*, 151.

48. Gobineau, *Inequality*, 208.

49. Gobineau, *Essai*, 301. "Certainement l'élément noir est indispensable pour développer le génie artistique dans une race, parce que nous avons vu quelle profusion de feu, de flammes, d'étincelles, d'entraînement, d'irréflexion réside dans son essence, et combien l'imagination, ce reflet de la sensualité, et toutes les appétitions vers la matière le rendent propre à subir les impressions que produisent les arts, dans un degré d'intensité tout à fait inconnu aux autres familles humaines."

50. Gobineau, *Inequality*, 209.

51. See Firmin, *Equality*, 1–14.

52. Firmin, *Equality*, 156.

53. Firmin, *Equality*, 106.

54. For his insistence that the statistics compiled by Broca and others frequently contradicted the conclusions they drew, see Firmin, *Equality*, 96–109.

55. Herbert Spencer, *Les bases de la morale évolutioniste* [The foundations of evolutionary ethics] (Paris: Librairie Bailliere & Co., 1881), 206, quoted in Firmin, *Equality*, 385.

56. Spencer used the phrase "survival of the fittest" in *The Principles of Biology* (1864; repr., New York: D. Appleton & Co., 1910), 264. Darwin appropriated the phrase in the fifth edition of *On the Origin of Species* (1869).

57. Firmin, *Equality*, 10. Firmin is referring to Topinard, *L'Anthropologie*, 2. Bertillon had written "divisions typiques (variété, race, espèce, s'il y a lieu)."

58. Firmin, *Equality*, 270.

59. Firmin, *Equality*, 156.

60. Clémence-Auguste Royer, preface to *De l'origine des espèces*, by Charles Darwin, trans. Clémence-Auguste Royer, 4th ed. (1862; repr., Paris: Guillaumin et Masson, 1885), lxi, quoted in Firmin, *Equality*, 271. On the disruptive nature of Royer's contributions to the Society, see Robert Bernasconi, "A Haitian in Paris: Anténor Firmin as a Philosopher Against Racism," *Patterns of Prejudice* 42, no. 4–5 (2008): 375–76. On the muted reception of Firmin's book in Paris, see Anne-Marie Drouin-Hans, "Hierarchy of Races, Hierarchy in Gender: Anténor Firmin and Clémence Royer," *Ludus Vitalis* 13, no. 23 (2005): 163–80.

61. Quoted in *Congrès international des sciences ethnographiques* (Paris, 1878), 196, quoted in Firmin, *Equality*, 205. The English translation of *Equality* uses "White" and "Negro" for Firmin's "blanc," "nègre," and "noir." On the abnormality of the métis in eighteenth- and nineteenth-century science, see Claude Blanckaert, "Of Monstrous Métis? Hybridity, Fear of Miscegenation, and Patriotism from Buffon to Paul Broca," in *The Color of Liberty: Histories of Race in France*, ed. Sue Peabody and Tyler Stovall (Durham, NC: Duke University Press, 2003), 42–70.

62. Firmin, *Equality*, 205.

63. Firmin, *Equality*, 270.

64. Firmin, *Equality*, 227, 231, 394. For a more recent and highly controversial version of the racial composition of ancient Egypt, see Martin Bernal, *Black Athena: The Afroasiatic Roots of Classical Civilization*, 3 vols. (1987; repr., New Brunswick, NJ: Rutgers University Press, 2006). On Bernal, see Jacques Berlinerblau, *Heresy in the University: The "Black Athena" Controversy and the Responsibility of American Intellectuals* (New Brunswick, NJ: Rutgers University Press, 1999). See also Gilles Boëtsch, "Noirs ou blancs: une histoire de l'anthropologie biologique de l'Égypt," *Égypte/Monde Arabe* 14 (1995): 113–38, https://doi.org/10.4000/ema.643. On Firmin and the African character of pharaonic Egypt, see Célucian L. Joseph, "Anténor Firmin, the 'Egyptian Question,' and Afrocentric Imagination," *Journal of Pan African Studies* 7, no. 2 (2014): 127–76.

65. Firmin, *Equality*, 398–99.

66. Firmin, *Equality*, lviii, 243, 244.

67. Firmin, *Equality*, 448, 428, 433, 434, 435.

68. For many years, only three versions of Firmin's book were available, the original and two Haitian editions, published respectively in 1968 (Panorama) and 1985 (Fardin). In 2003, a French edition appeared with an introduction by Ghislaine Géloin (Paris: L'Harmattan), and in 2022, an independently published print-on-demand edition became available. On the flaws both in the original text and in the translation, as well as reflections on Firmin's place in the contemporary discourse on race, see Bernasconi, "A Haitian in Paris."

69. Firmin, *Equality*, 401.

70. Broca, Mémoires, 3:245.

71. See Wartelle, "La Société d'Anthropologie de Paris de 1859 à 1920," 50. The Paris Anthropology Society was not the only one to experience a crisis of identity over issues of scientific style. On the tumultuous self-repudiating history of the London Anthropological Institute—which had been created in the wake of the demise of the Anthropological Society, which had itself evolved out of the Aborigines Protection Society—see George W. Stocking Jr., "What's in a Name? The Origins of the Royal Anthropological Institute (1837–71)," *Man* 6, no. 3 (September 1971): 369–90.

72. Topinard, *L'Anthropologie*, 20, 193, 511.

73. Paul Topinard, *Éléments d'Anthropologie Générale* (Paris: Adrien Delahaye and Émil Lescrosniers, 1885), 78.

74. Paul Topinard, "On 'Race' in Anthropology" ["De la race en anthropologie"] (paper presented at the Congrés international d'archeologie préhistorique et d'anthropologie, 11th session, Moscow, 1892), in *This Is Race: An Anthology Selected from the International Literature on the Races of Man*, ed. Earl W. Count (New York: Henry Schuman, 1950), 161–70, 171. Topinard did reserve a certain legitimacy for the term race as applying to "the high sphere of general types," but said that these types, like Gobineau's Aryans, had long since vanished from the earth. Topinard was far from the first to register skepticism about the suitability of race for scientific inquiry. In 1850, Latham wrote that "a race is a class of individuals concerning which there are doubts as to whether they constitute a separate species or a variety of a recognized

one. Hence the term is *subjective*." Robert Gordon Latham, *The Natural History of the Varieties of Man* (London: Van Voorst, 1850), 564.

75. Charles Darwin, letter to Asa Gray, in *The Correspondence of Charles Darwin*, ed. Frederick Burkhardt and Sydney Smith (Cambridge: Cambridge University Press, 1997), 10:241.

76. See Joy Harvey, *Almost a Man of Genius: Clémence Royer, Feminism and Nineteenth-Century Science* (New Brunswick, NJ: Rutgers University Press, 1997).

Chapter Five: The Fall and Rise of Race

1. Jean-Claude Wartelle, "La Société d'Anthropologie de Paris à 1859–1920," *Revue d'Histoire des Sciences Humaines*, 1, no. 10 (2004): 125–71, https://www.cairn.info/revue-histoire-des-sciences-humaines-2004-1-page-125.htm.

2. Anthropologists, according to Emmanuelle Sibeud, "were at a loss to offer convincing support to colonial and metropolitan authorities," and the discipline came to seem "a peripheral and eventually troublesome form of leisure, implicitly reserved to colonial physicians after World War I." Emmanuelle Sibeud, "A Useless Colonial Science? Practicing Anthropology in the French Colonial Era, 1880–1960," *Current Anthropology* 53, supp. 5 (2012): S83, S93, https://www.journals.uchicago.edu/doi/epdf/10.1086/662682.

3. Jean-Louis Armand de Quatrefages de Bréau, *The Human Species* (New York: Appleton & Company, 1879), 154–67, originally published as *L'Espèce humaine* (1877). For the source of many arguments in this book, see Jean-Louis Armand Quatrefages de Bréau, *The Prussian State, Ethnologically Considered*, trans. Isabella Innes (London: Virtue & Co., 1872). On the accumulation of doubts about methods and premises in the Anthropology Society of Paris, see Peter Cryle and Elizabeth Stephens, *Normality: A Critical Genealogy* (Chicago: University of Chicago Press, 2017), esp. 142–79.

4. Quatrefages, *Human Species*, 83.

5. Quatrefages, *Human Species*, 31.

6. Quatrefages, *Human Species*, 244, 251, 259.

7. Quatrefages, *Human Species*, 283.

8. Quatrefages, *Human Species*, 255.

9. Quatrefages, *Human Species*, 261, 262.

10. Rudolf Virchow, "Heredity and the Formation of Races," in *This Is Race: An Anthology Selected from the International Literature on the Races of Man*, ed. Earl W. Count (New York: Henry Schuman, 1950), 185. Originally "Rassenbildung und Erblichkeit," in *Festschrift für Adolf Bastian* (Berlin, 1896), 1–43; this essay was translated for this publication.

11. See Andrew Zimmerman, "Anti-Semitism as Skill: Rudolf Virchow's *Schulstatistik* and the Racial Composition of Germany," *Central European History* 32, no. 4 (2008): 409–29, https://doi.org/10.1017/S0008938900021762. Replying to the contention of Quatrefages in 1870 that the German people were not Teutons at all but rather Slavo-Finns, Virchow argued, largely on the basis of data from millions of skull measurements, that modern Germans were racially mixed.

12. The reference to Nordic mysticism was intended as a criticism of Ernst Haeckel and others. The term *la race nordique* would also be used by Joseph Deniker in *The Races of Man*, and would be taken up by Madison Grant in *The Passing of the Great Race* (1916).

13. Virchow, "Heredity," 189.

14. Joseph Deniker, *The Races of Man: An Outline of Anthropology and Ethnography* [*Les races et les peoples de la terre*] (London: Walter Scott, Limited, 1900).

15. Deniker himself had examined eleven thousand crania, but this was a tiny fraction of the number that had been accumulated. William Z. Ripley estimated that over one million adults and ten million children in the United States and Europe had been measured with the intention to determine their racial identity. William Z. Ripley, *The Races of Europe: A Sociological Study* (New York: D. Appleton, 1899), 58–77. In 1900, Gossett reports, one particularly dedicated researcher had measured a single skull in five thousand different ways. Thomas F. Gossett, *Race: The History of an Idea in America* (1963; repr., New York: Oxford University Press, 1997), 76. In the 1920s, Lancelot Hogben described the vast proliferation of such measurements as an instance of the "stupendous persistence in fruitless and trivial exploits of repetitive mensuration" in racial science. Quoted in Nancy Stepan, *The Idea of Race in Science: Great Britain, 1800–1960* (Houndmills, UK: Macmillan, 1982), xviii.

16. Deniker, *Races of Man*, 2.

17. Deniker, *Races of Man*, 3, 4.

18. Walter Scheidt, "The Concept of Race in Anthropology and the Divisions into Human Races from Linnaeus to Deniker," in *This Is Race: An Anthology Selected from the International Literature on the Races of Man*, ed. Earl W. Count (New York: Henry Schuman, 1950), 354–91, 389, originally published as *Beiträge zur Geschichte der Anthropologie* (1924–25).

19. Charles Darwin, *The Descent of Man, and Selection in Relation to Sex* (London: John Murray, 1871), 3.

20. Henry Adams, *The Education of Henry Adams* (1918; repr., New York: Modern Library, 1931), 411–12.

21. Franz Boas, "History and Science in Anthropology: A Reply," *American Anthropologist* 38, no. 1 (1936): 140.

22. Franz Boas, *Changes in Bodily Form of Descendants of Immigrants* (1912; repr., Boston: Adamant Media Corporation, 2006), 7. By insisting on culture or environment as opposed to biology, Boas reoriented and professionalized the field of anthropology. In doing so, he produced other effects. In the judgment of Gossett (*Race*, 418), "It is possible that Boas did more to combat race prejudice than any other person in history."On Boas and the approach to race within the discipline of anthropology, see Rachel Caspari, "Race, Then and Now: 1918 Revisited," *American Journal of Physical Anthropology* 165 (2018): 924–38, https://doi. org/10.1002/ajpa.23417. For a good selection of Boas's texts, see George W. Stocking Jr., ed., *A Franz Boas Reader: The Shaping of American Anthropology, 1883–1911* (Chicago: University of Chicago Press, 1974). Even more striking, perhaps, than Boas's discovery about the skull measurements of immigrants is the fact that this possibility was raised by Samuel George Morton's craniological measurements, and was being seriously proposed in the mid-nineteenth century by Charles Hamilton Smith and Hunt as well as

Broca. See James Hunt, *The Negro's Place in Nature: A Paper Read Before the London Anthropological Society* (New York: Van Evrie, Horton & Company, 1866), 32. For a different view of Boas on race, however, see Lee D. Baker, "The Racist Anti-Racism of American Anthropology," *Transforming Anthropology* 29, no. 2 (October 2021): 127–42. This article focuses on the address given—at the invitation of Du Bois—by Boas at the 1910 meeting of the National Association for the Advancement of Colored People on "The Real Race Problem" in which he encouraged his audience to assimilate into the larger American culture, thereby facilitating "the gradual process of lightening up this large body of people by the influx of white blood" (127).

23. In 1881, Gustav Le Bon complained that "current anthropology is this accessory branch of comparative osteology devoted to the study of variations in the cranium among human races." Gustav Le Bon, "L'Anthropologie actuelle et l'étude des races," *Revue Scientifique* 28 (1881): 772.

24. The sixty texts collected in Earl W. Count, ed., *This Is Race: An Anthology Selected from the International Literature on the Races of Man* (New York: Henry Schuman, 1950) constitute a remarkable archive of scientific research on the subject, much of it translated for the first and only time. All the terms and quotes mentioned in this paragraph and the preceding one are taken from texts in this volume that were written between 1900 and 1939.

25. For "types less dowered," see Roland B. Dixon, *The Racial History of Man* (New York: Charles Scribner's Sons, 1923), 516.

26. Pearson explained his "scientific" view of a nation as an "organized whole, kept up to a high pitch of internal efficiency by insuring that its numbers are substantially recruited from the better stocks, and kept up to a high pitch of external efficiency by contest, chiefly by way of war with inferior races." Karl Pearson, *National Life from the Standpoint of Science* (London: Adam and Charles Black, 1901), 43–44. An atheist, freethinker, and socialist, Pearson died in 1936.

27. The 1911 entry was written by Thomas Athol Joyce, a British ethnographer, and Walter Francis Willcox, a statistician working at the US Census Bureau. *Encyclopedia Britannica*, 11th ed., https://en.wikisource.org/wiki/1911_Encyclop%C3%A6dia_Britannica/Negro. The 1922 entry was the work of Willcox alone. *Encyclopedia Britannica*, 12th ed., https://en.wikisource.org/wiki/1922_Encyclop%C3%A6dia_Britannica/Negro.

28. Ruth Benedict, "Racism: The 'Ism' of the Modern World," in *Race: Science and Politics* (1940; repr., Athens: University of Georgia Press, 2019), 3–8. Benedict begins this essay, which virtually invented the term racism, by citing a French pro-Aryan, Vacher de Lapouge, who wrote in the late 1880s, "I am convinced that in the next century millions will cut each other's throats because of 1 or 2 degrees more or less of cephalic index." She wrote the text in 1939.

29. Ruth Benedict, *Race: Science and Politics* (1940; repr., Athens: University of Georgia Press, 2019), 150.

30. Ashley Montagu, *Man's Most Dangerous Myth: The Fallacy of Race* (New York: Columbia University Press, 1942). In a later statement, Montagu described race as "the phlogiston of our time." Ashley Montagu, "Introduction," *The Concept of Race*, ed. Ashley Montagu (New York: Free Press, 1964), xii. Montagu was one of the ten scientists invited to serve on a panel that produced a widely circulated and

controversial statement in 1950 on the subject of race—the first of four that would ultimately be issued. See UNESCO, *The Race Concept: Rèsults of an Inquiry* (Paris: UNESCO, 1952). This publication included an initial five-page statement composed by Montagu and others that affirmed principles of equality and the "brotherhood of man," followed by eighty pages of comments, criticisms, and objections to the statement by others. For a good introduction to this episode, see Ivan Hannaford, *Race: The History of an Idea in the West* (Baltimore: Johns Hopkins University Press, 1996), 385–92. For a comprehensive account of the collapse of scientific support for race as a biological fact, see Elazar Barkan, *The Retreat of Scientific Racism: Changing Concepts of Race in Britain and the United States Between the World Wars* (Cambridge: Cambridge University Press, 1992).

31. On race as "mental tool," see Staffan Müller-Wille, "Race and History: Comments from an Epistemological Point of View," *Science, Technology, and Human Values* 39, no. 4 (2014): 597–606. On "built information environment," see Geoffrey C. Bowker and Susan Leigh Star, *Sorting Things Out: Classification and Its Consequences*, (Cambridge, MA: MIT Press, 1999), 5.

32. W.E.B. Du Bois, "The Conservation of Races," in *W.E.B. Du Bois: Writings*, ed. Nathan Huggins (New York: Library of America, 1986), 820, 817.

33. Du Bois, "Conservation of Races," 820, 825.

34. Du Bois, "Conservation of Races," 817, 816.

35. According to David Levering Lewis, "Just exactly what race was Du Bois never did manage to say clearly." David Levering Lewis, *W. E. B. Du Bois: Biography of a Race, 1868–1919* (New York: Henry Holt and Company, 1993), 171.

36. Kwame Anthony Appiah, "The Uncompleted Argument: Du Bois and the Illusion of Race," in *"Race," Writing, and Difference*, ed. Henry Louis Gates Jr. and Kwame Anthony Appiah (Chicago: University of Chicago Press, 1986), 36, 134.

37. Appiah, "Uncompleted Argument," 36.

38. Kwame Anthony Appiah, *Lines of Descent: W. E. B. Du Bois and the Emergence of Identity* (Cambridge, MA: Harvard University Press, 2014), 200n7. For other perspectives on Du Bois, see Paul Gilroy, *The Black Atlantic: Modernity and Double Consciousness* (Cambridge, MA: Harvard University Press, 1995), esp. chap. 4, which centers on Du Bois. See also Simon Gikandi, "W. E. B. Du Bois and the Identity of Africa," *Gefame: Journal of African Studies* 2, no. 1 (2005), https://quod.lib.umich.edu/g/gefame/4761563.0002.101/—w-e-b-dubois-and-the-identity-of-africa. The dialectic enacted between Du Bois and Appiah about race is provocatively reframed in Walter Benn Michaels, *The Trouble with Diversity: How We Learned to Love Identity and Ignore Inequality* (New York: Metropolitan, 2015).

39. Although the "social construction" theory dominates many recent discussions, Nell Irvin Painter begins her book *The History of White People* (New York: W. W. Norton & Company, 2010) with the statement that "today we think of race as a matter of biology" (xi). A good deal of recent research has focused on ways of determining average genetic differences among human populations—differences that are generally far smaller than those that exist among individuals, but can still be useful in, for example, predicting risk factors for various diseases. The use of race as an independent variable is fairly widespread in the United States, though less so in other countries, notably France. For a sophisticated defense of "racial" naturalism in the generation of

meaningful statistics in certain medical contexts, see Ian Hacking, "Why Race Still Matters," *Daedalus* (Winter 2005), https://www.amacad.org/publication/daedalus/why-race-still-matters). Hacking's instances involve "populations" such as "West African" that are not, however, "races." Recent genetic science has also avoided the term race, with scientists sometimes placing the word in quotation marks, and generally preferring "ethnicities" or "genetic clusters" that do not map easily onto the kinds of traits traditionally identified as racial. See Noah A. Rosenberg, Jonathan K. Pritchard, James L. Weber, Howard M. Cann, Kenneth K. Kidd, Lev A. Zhivotovsky, and Marcus W. Feldman, "Genetic Structure of Human Populations," *Science* 298, no. 5602 (December 20, 2002): 2381–85, https://www.science.org/doi/10.1126/science.1078311. Australian anthropologist Adam Hochman has argued that race has an inescapably biological implication. Describing race as "a biological concept which fails to refer," Hochman proposes as an alternative an "interactive constructionism" that unlike "social constructionism," rejects any biological component to race. The premise of interactive constructionism is that "there are no races, only racialized groups—groups mistakenly believed to be races." See Adam Hochman, "Replacing Race: Interactive Constructionism about Racialized Groups," Brewminute, https://brewminate.com/replacing-race-interactive-constructionism-about-racialized-groups/. See also Adam Hochman, "Has Social Construction About Race Outlived Its Usefulness? Perspectives from a Race Skeptic," *Biology and Philosophy* 36, no. 6 (2022): 1–22, https://philpapers.org/rec/HOCHSC; Adam Hochman, *The Race Illusion: On the Reality of Racialization and the Myth of Race* (Oxford: Oxford University Press, 2025). For a complete listing of Hochman's contributions to the subject, see https://philpeople.org/profiles/adam-hochman.

40. For an account of this incident, see Lewis, *W. E. B. Du Bois*, 373.

41. Robert Wald Sussman, *The Myth of Race: The Troubling Persistence of an Unscientific Idea* (Cambridge, MA: Harvard University Press, 2014), 307.

42. For a recent argument in favor of "abandoning the delusion of race," see Subrema E. Smith and David Livingstone Smith, "The Trouble with Race and Its Many Shades of Deceit," *New Lines Magazine*, March 29, 2023, https://newlinesmag.com/argument/the-trouble-with-race-and-its-many-shades-of-deceit/. See also the now-classic work by Barbara and Karen Fields, *Racecraft: The Soul of Inequality in American Life* (New York: Verso, 2012), which compares races to witches as imagined evils, describes the concept of race as "the principal unit and core concept of *racism*," and argues that racial science should be called "bioracism" (17). Index

INDEX

A NOTE ON THE TYPE

THIS BOOK has been composed in Miller, a Scotch Roman typeface designed by Matthew Carter and first released by Font Bureau in 1997. It resembles Monticello, the typeface developed for The Papers of Thomas Jefferson in the 1940s by C. H. Griffith and P. J. Conkwright and reinterpreted in digital form by Carter in 2003.

Pleasant Jefferson ("P. J.") Conkwright (1905–1986) was Typographer at Princeton University Press from 1939 to 1970. He was an acclaimed book designer and AIGA Medalist.

The ornament used throughout this book was designed by Pierre Simon Fournier (1712–1768) and was a favorite of Conkwright's, used in his design of the *Princeton University Library Chronicle.*